98-WSM-2P

A Peer Review of the City of Lincoln, Nebraska Salt Creek Site-Specific Ammonia Water Quality Criteria

by:

Cynthia Paulson, Ph.D., P.E.
Bret Linenfelser
Brown and Caldwell

Michael Bastian
CH2M Hill

Lyle Christensen
City of Lincoln Wastewater Division

William Rue
EA Engineering, Science, and Technology

Robert Weaver
Kelly & Weaver

2006

WERF

The Water Environment Research Foundation, a not-for-profit organization, funds and manages water quality research for its subscribers through a diverse public-private partnership between municipal utilities, corporations, academia, industry, and the federal government. WERF subscribers include municipal and regional water and wastewater utilities, industrial corporations, environmental engineering firms, and others that share a commitment to cost-effective water quality solutions. WERF is dedicated to advancing science and technology addressing water quality issues as they impact water resources, the atmosphere, the lands, and quality of life.

For more information, contact:
Water Environment Research Foundation
635 Slaters Lane, Suite 300
Alexandria, VA 22314-1177
Tel: (703) 684-2470
Fax: (703) 299-0742
www.werf.org
werf@werf.org

This report was co-published by the following organization. For non-subscriber sales information, contact:

IWA Publishing
Alliance House, 12 Caxton Street
London SW1H 0QS, United Kingdom
Tel: +44 (0) 20 7654 5500
Fax: +44 (0) 20 7654 5555
www.iwapublishing.com
publications@iwap.co.uk

Library of Congress Catalog Card Number: 2006937497
IWAP ISBN: 978-1-84339-775-5

Brown and Caldwell, CH2M Hill, City of Lincoln Wastewater Division, EA Engineering, Science, and Technology, Kelly & Weaver

ACKNOWLEDGMENTS

The project team would like to acknowledge the contributions of many members of the City of Lincoln Wastewater Division that were very involved with the management and focus of the project, including Gary Brandt, Lyle Christensen, Steve Masters, and Randy Wilson; Steve Huggenburger as the City's attorney; and many Wastewater Division staff and others from the Public Works and Utilities Department who performed much of the field and laboratory work associated with the *in situ* testing. Throughout the project, the Wastewater Division worked closely with regulators from EPA Region 7 and the Nebraska Department of Environmental Quality as well as other stakeholders, such as the Nebraska Game & Parks Commission, the U.S. Geological Survey, the Lower South Platte Natural Resources District, the Nebraska Department of Water Resources, and the University of Nebraska, who are recognized for their input. Project support and encouragement by Public Works and Utilities Department Directors, Dick Erixson and Allan Abbott were critical to the work and its application.

Report Preparation

Project Team:

Michael Bastian
CH2M Hill

Lyle Christensen
City of Lincoln Wastewater Division

Bret Linenfelser
Brown and Caldwell

Cynthia Paulson, Ph.D., P.E.
Brown and Caldwell

William Rue
EA Engineering, Science, and Technology

Robert Weaver
Kelly & Weaver

Peer Review Committee

Scott Belanger, Ph.D.
Procter & Gamble

Robert Berger
Consultant

Jerry Diamond, Ph.D.
Tetra Tech, Inc.

Norm LeBlanc
Hampton Roads Sanitation District

James Karr, Ph.D.
University of Washington

Thomas Sprehe
KCI Technologies, Inc.

Water Environment Research Foundation Staff

Director of Research: Daniel M. Woltering, Ph.D.

ABSTRACT AND BENEFITS

Abstract:

This report is the summary of a portion of a larger-scale project titled *Salt Creek Water Quality Studies* (SCWQS) performed by the City of Lincoln, Nebraska, for an approximate seven-year period from 1994 through 2000, to address proposed ammonia effluent limits for the city's two wastewater treatment plants (WWTPs). The project focused on characterizing Salt Creek by quantifying the existing biological community, evaluating existing water quality, physical and hydrologic conditions, and evaluating the impact of ammonia discharges from the WWTPs to the biological community compared to other stressors. The portion of the project that is the primary focus of this report is the *in situ* toxicity study performed in Salt Creek that modeled typical laboratory tests that introduce a species of fish to known levels of ammonia to determine a dose-response relationship. The *in situ* toxicity study results would be the basis for developing a site-specific chronic ammonia water quality criterion for Segment LP2-20000 of Salt Creek. The Water Environment Research Foundation (WERF) provided peer review for this project, which resulted in credible research results that were used to develop alternative ammonia discharge limits for the City's WWTPs. Substantial support was also provided by the Nebraska Department of Environmental Quality and the Region VII Environmental Protection Agency.

Benefits:

- Characterizes levels and variability of chemical parameters of concern that occur during the exposure of ammonia to selected fish species.
- Links ammonia levels to sublethal (chronic) responses of the selected fish species.
- Develops relationships between measured ammonia levels and response of exposed fish.
- Links fish response to identified human activities on Salt Creek.
- Determines appropriate levels of ammonia for the protection of local aquatic communities.
- Develops a technical basis for calculating site-specific ammonia criteria and resulting effluent limits.
- Complies with the Wastewater Division's mission for wastewater treatment and potential impacts to rate payers.

Keywords: Water Quality; Site-Specific Criteria; *In-Situ* Toxicity Testing; Permitting; Ammonia

TABLE OF CONTENTS

LIST OF TABLES

LIST OF FIGURES

LIST OF ACRONYMS

CCC	Criteria Continuous Concentration
cfs	cubic feet per second
City	City of Lincoln, Nebraska
CV	coefficient of variation
GMCV	Genus Mean Chronic Value
IBI	Index of Biotic Integrity
ICI	Invertebrate Community Index
mg/L	milligrams per liter
NDEQ	Nebraska Department of Environmental Quality
NPDES	National Pollutant Discharge Elimination System
NSWQS	Nebraska Surface Water Quality Standards
ppt	parts per thousand
PVC	polyvinyl chloride
QA/QC	quality assurance/quality control
RM	river mile
SCWQS	Salt Creek Water Quality Studies
s.u.	standard units
UAA	use attainability analysis
U.S. EPA	U.S. Environmental Protection Agency
WERF	Water Environment Research Foundation
WWTP	wastewater treatment plants

EXECUTIVE SUMMARY

The City of Lincoln, Nebraska, (City) is approximately 65 square miles in size and is bordered on all sides by the unincorporated area of Lancaster County. At the time of the Salt Creek Water Quality Studies (SCWQS) the City had a population of 191,972. The City's two Wastewater Treatment Plants (WWTPs), Theresa Street and Northeast, provide wastewater treatment to industry and development within the City as well as portions of the surrounding County. The Theresa Street WWTP has a rated treatment capacity of 21 million gallons per day (mgd) and the Northeast WWTP has a rated capacity of 8 mgd. The WWTPs discharge directly into Segment LP2-20000 of Salt Creek, which is classified as Warmwater Habitat A. To meet the ammonia criteria associated with this classification, the WWTPs were assigned draft effluent permit limits as shown in the table below.

Proposed NPDES Chronic Ammonia Effluent Limits – Theresa Street and Northeast WWTPs.

Facility	30-Day Average Ammonia Effluent Limit by Season	
	Summer (March – October)	Winter (November – February)
Theresa Street WWTP	1.48 mg/L	2.70 mg/L
Northeast WWTP	1.54 mg/L	2.70 mg/L

Salt Creek Segment LP2-20000 is channelized for flood protection, and over 100 storm water outfalls discharge runoff to the stream. Segment LP2-20000 of Salt Creek is also heavily influenced by the underlying Dakota Sandstone Formation near the upstream portion of the Segment. Sodium chloride from the formation dissolves in the groundwater contributing to the stream, and chloride concentrations in the range of 1,000 to 2,000 milligrams per liter (mg/L) are common during median flow conditions. In parts of the Segment LP2-20000 reach salinity increases to 10 to 15 times higher than ambient levels due to ground water flows through a natural salt formation at the upstream end of the segment. Salt Creek has extremely variable flows which are dominated by storm water runoff during large precipitation events. Segment LP2-20000 is also affected by agricultural runoff, urban storm water runoff, and industrial discharges.

Given the unique characteristics of Salt Creek Segment LP2-20000, the City was concerned that meeting the proposed ammonia effluent limits would not result in improved aquatic life in Salt Creek. Based on historical chemical and biological data there seemed to be a disconnect between the biological, physical and chemical state of the Salt Creek, with its channelized features and high salinity, and the ammonia limits being suggested to protect aquatic life. The City began a program to characterize the biology, habitat, and chemistry of Salt Creek to determine if the proposed ammonia discharge limits would actually improve the aquatic life in the creek or if other physical or chemical factors would be the limiting factor in Salt Creek.

From June 1994 through October 1999, six summer season and five winter season biological assessments were conducted on Segment LP2-20000 of Salt Creek in both impacted and reference sites. The biological assessments focused on the collection of fish and benthic macroinvertebrates, characterization of water quality, and habitat assessment.

Results of the biological assessments, including characterization of Salt Creek water quality and habitat, indicated that there is no relationship between the fish community and benthic macroinvertebrate metrics and ammonia in Salt Creek. In addition, habitat quality was judged to be poor throughout the reach of Segment LP2-20000, from the channelized and saline reach of Salt Creek above the WWTPs through the most downstream station, because of the predominance of uniform depth and velocity; homogenous, unstable sand substrate; the lack of pools and cover; and elevated salinity concentrations.

In summary, the key findings of the Salt Creek studies were:

- The Dakota sandstone formation is the predominant factor affecting water quality within the Segment LP2-20000 drainage through the addition of chloride.
- Flood control channelization has fundamentally altered the habitat in Salt Creek.
- Based on habitat quality, the Index of Biotic Integrity (IBI) index scores were less than predicted at the reference sites and greater than predicted at the ammonia-impacted sites.
- Channelization, chloride, temperature, and reservoir stocking are key determinant factors controlling biological communities in Salt Creek.
- Ammonia and several other constituents are not determinant factors for Salt Creek biological communities.
- The aquatic life use in Segment LP2-20000 is limited by naturally occurring chloride concentrations, and by human caused hydrologic modifications.
- The Class A Warmwater aquatic life use is not presently existing and cannot be attained in Segment LP2-20000 of Salt Creek.
- A site-specific application of water quality data is appropriate to develop site-specific Water Quality Standards.
- There is an inconsistency or disconnect between instream biological assessment data and national chemical specific data and criteria for ammonia. This disconnect is likely caused by many factors including the mitigative effects of salinity on ammonia toxicity, varying exposure, genetic adaptation, and the tolerance of the indigenous community.

From these key findings, the City determined that site-specific criteria for ammonia would be most appropriate for Segment LP2-20000 of Salt Creek. The City recognized that in order to develop site-specific criteria for ammonia the technical approach needed to be scientifically sound and defensible and fit in to the current regulatory framework. The City proposed the SCWQS project for WERF peer review, expecting that such a review would make the approach to developing a site-specific ammonia criterion robust and defensible and would provide an independent perspective. WERF Peer Review members attended two workshops with the project team to review the findings to date and develop recommendations.

Based on the WERF Peer Review Panel recommendations, the City of Lincoln initiated *in situ* testing in Salt Creek to more accurately characterize the complex interactions of ammonia in Salt Creek waters and support the development of a site-specific chronic ammonia criterion. The *in situ* toxicity test consisted of caged fish that were continuously exposed for 15-day and 30-day periods at nine in-stream locations. The *in situ* program was developed by the City and their consulting team with technical input from the WERF Peer Review Panel. Technical and regulatory support for the development of the *in situ* study program was also provided by the

NDEQ, Surface Water Section, Water Quality Division, and U.S. EPA Region VII, Water Quality Management Branch.

The selected fish species for *in situ* testing needed to be representative of the fish community in Salt Creek, sensitive to ammonia, able to tolerate ambient Salt Creek salinities, and behaviorally adaptable to being held in a cage or test chamber. Both channel catfish and fathead minnow were chosen to meet these requirements. In order to prepare for the actual *in situ* study, a pilot study was performed to confirm the suitability of in situ testing to measure toxicity. The *in situ* test itself was run for 30 days, with half of the fish cages removed and analyzed at 15 days and the others maintained for the full 30 days.

In order to identify an impact threshold for the aquatic biota in Salt Creek, the ammonia exposure concentration that affected 20 percent of the fish population, or IC20, was calculated for the *in situ* study. The results of the test provided both a 15- and 30-day IC20 value for channel catfish and fathead minnow. The 30-day IC20 value for the channel catfish was calculated to be 3.85 mg/L based on total biomass. The corresponding 15-day IC20 value was calculated as 7.33 mg/L, which is within the relative range expected for a shorter duration test. Since there was less than a 20 percent decrease in fathead minnow biomass at downstream stations compared to the control at the highest 30-day average in-stream exposure concentration, the 30-day IC20 values for the fathead minnow were greater than 9.98 mg/L. These *in situ* test results were then used to adjust the ammonia criteria for Salt Creek Segment LP2-20000, resulting in a 28% increase in ammonia chronic criteria.

CHAPTER 1.0

SALT CREEK WATER QUALITY STUDIES BACKGROUND

The City of Lincoln, Nebraska, (City) is approximately 65 square miles in size and is bordered on all sides by the unincorporated area of Lancaster County. At the time of the Salt Creek Water Quality Studies (SCWQS) the City had a population of 191,972 with an anticipated growth rate of approximately one percent per year. The local economy is driven by the abundance of commercial and industrial activities within City limits, including electronics, manufacturing, and a variety of agricultural and transportation related industries. Agriculture is also very prominent. In the area surrounding Lincoln, large tracts of land are being used to support a variety of agricultural practices.

The City's two Wastewater Treatment Plants (WWTPs), Theresa Street and Northeast, provide wastewater treatment to industry and development within the City as well as portions of the surrounding County. The WWTPs discharge directly into Segment LP2-20000 of Salt Creek. The Theresa Street WWTP has a rated treatment capacity of 21 million gallons per day (mgd) and the Northeast WWTP has a rated capacity of 8 mgd.

Salt Creek is a fourth-order tributary to the Platte River in southeast Nebraska. For 21.6 miles, Segment LP2-20000 flows through the City of Lincoln and rural agricultural land in Lancaster County and extends from the confluence with Beal Slough to the confluence with Rock Creek. The stream is channelized for flood protection, and over 100 storm water outfalls discharge runoff to the stream. In addition, the City's Theresa Street WWTP and Northeast WWTP discharge into the Segment and contribute a significant portion of the total stream flow under median and low-flow conditions. The historical (1967 – 1999) median Salt Creek flow above the Theresa St. WWTP for the summer regulatory season (March – October) is 94.7 cubic feet per second (cfs), while the average summer season Theresa St. WWTP effluent flow rate is 31.4 cfs.

Segment LP2-20000 of Salt Creek is also heavily influenced by the underlying Dakota Sandstone Formation near the upstream portion of the Segment. Sodium chloride from the formation dissolves in the groundwater contributing to the stream, and chloride concentrations in the range of 1,000 to 2,000 milligrams per liter (mg/L) are common during median flow conditions. In parts of the Segment LP2-20000 reach salinity increases to 10 to 15 times higher than ambient levels due to ground water flows through a natural salt formation at the upstream end of the segment. Salt Creek has extremely variable flows which are dominated by stormwater runoff during large precipitation events. Segment LP2-20000 is also affected by agricultural runoff, urban storm water runoff, and industrial discharges (SCWQSR 1996 V.1 of 2.).

Due to the multiple natural impacts to Segment LP2-20000 of Salt Creek, a use attainability analysis (UAA) was conducted in 1982 by the Nebraska Department of Environmental Quality (NDEQ). The UAA states in part: "Because of physical habitat limitations due to channelization, Segment LP-3a [now LP2-20000] of Salt Creek appears capable of maintaining only a limited warm-water fishery" (NDEQ 1982*)*. At the time of the UAA, the Salt Creek designated use was Limited Warmwater Habitat under the Nebraska Surface Water Quality Standards (NSWQS). In 1991 Segment LP3a of Salt Creek was changed to Segment LP2-20000 and redesignated the "highest" aquatic life use, —Warmwater Habitat A, and reclassified for the following uses:

- Primary Contact Recreation
- Aquatic Life Warmwater Class A
- Water Supply Agricultural B
- Aesthetics

Incorporated in the Aquatic Life Warmwater Class A designation are site-specific water quality criteria for un-ionized ammonia.

Based on these changes the City did not believe that the change from a Limited Warmwater designation to a Warmwater Class A designation was well justified, particularly considering there was no change in the factors limiting the use (e.g., channelization and related limiting habitat and high salinity) in Salt Creek from 1982, when the UAA identified these limiting factors, to 1991. No follow-up UAA was performed for the 1991 reclassification.

1.1 Need for Salt Creek Water Quality Studies

At the time Salt Creek was reclassified, and ammonia criteria were implemented, the City's two WWTPs were operating under National Pollutant Discharge Elimination System (NPDES) permits, which did not contain effluent limits for ammonia. Typical WWTP ammonia discharge concentrations were in the range of 15 to 20 mg/L for both the Theresa Street WWTP and the Northeast WWTP. In 1994, during the NPDES permit renewal process, the City received draft NPDES permits from NDEQ that included ammonia effluent limits based on the Warmwater Habitat A beneficial use. The proposed NPDES chronic ammonia permit limits, and associated chronic criteria, are shown in Table 1-1.

Table 1-1. Proposed NPDES Chronic Ammonia Effluent Limits – Theresa Street and Northeast WWTPs.

	30-Day Average Ammonia Effluent Limit by Season	
Facility	Summer (March – October)	Winter (November – February)
Theresa Street WWTP	1.48 mg/L	2.70 mg/L
Northeast WWTP	1.54 mg/L	2.70 mg/L

Based on the historical level of ammonia treatment at both the Theresa Street and Northeast WWTPs, it would not be possible for the existing WWTPs to consistently meet the proposed ammonia effluent limits, especially the chronic (30-day average) limits.

1.1.1 Required Level of Ammonia Removal

In order to meet the proposed ammonia effluent limits, the City evaluated required improvements at both the Theresa Street and Northeast WWTPs and associated costs to achieve compliance with the proposed limits. Anticipating the need for complete ammonia removal, it was estimated that the total capital cost to upgrade both WWTPs would be approximately $35 million. This amount would be in addition to approximately $170 million to be spent over

20 years in planned wastewater facility construction needs identified in the City's Capital Improvement Program.

1.1.2 Concerns with Required Level of Ammonia Removal

Given the unique characteristics of Salt Creek Segment LP2-20000, the City was concerned that meeting the proposed ammonia effluent limits would not result in improved aquatic life in Salt Creek. Based on historical chemical and biological data there seemed to be a disconnect between the biological, physical and chemical state of the Salt Creek, with its channelized features and high salinity, and the ammonia limits being suggested to protect aquatic life. The mission for wastewater treatment at the Wastewater Division is

> *to provide a ". . continuing and comprehensive effort to assure the proper collection and treatment of current and future wastewater flows and loads in an environmentally sound and cost effective manner that protects human health and environment ."*

Given this dual mission, the City felt that it was their responsibility to ensure that if improvements to the wastewater treatment process were required, that they were justified. Since there was concern regarding the actual expected improvements in the aquatic biota in Salt Creek as a result of decreased ammonia discharges from the City's WWTPs, the City wanted to investigate alternative, and possibly more environmentally effective and cost effective, ways of getting the same or better results as compared to solely doing wastewater treatment plant upgrades.

CHAPTER 2.0

OVERVIEW OF SALT CREEK WATER QUALITY STUDIES

The SCWQS were completed to provide a sound scientific basis for the development of appropriate NPDES permit limits for the City's WWTPs, particularly for ammonia. Specifically, the studies addressed the following objectives:

- Comprehensive, integrated characterization of site-specific physical, chemical, and biological habitat conditions, particularly for the evaluation of the Warmwater Habitat A use designation and associated ammonia criteria.
- Development of a baseline water quality and flow database, which defines the effects of ammonia discharges on Salt Creek, including ammonia concentrations, dilution, decay, toxicity, and variation with flow.
- Evaluation of the relative importance of toxic effects from ammonia versus other water quality constituents and comparison of low-flow and storm runoff quality conditions.
- Preliminary assessment of Salt Creek compliance with water quality standards through the comparison of measured constituents to NDEQ water quality criteria.
- Support of long-term comprehensive watershed management planning for the Salt Creek Basin, including development of appropriate water quality effluent limits that consider the unique nature of Salt Creek and ensure the cost-effective protection of beneficial uses.
- Evaluation of effluent management options in parallel with water quality studies to assess the viability of various effluent management alternatives in conjunction with various ammonia permit limit scenarios to achieve compliance with future ammonia permit limits.

2.1 Study Approach

Sample site selection for the Salt Creek studies was based on temporal and spatial factors, as well as seasonal variation in chemical quality, physical habitat, and biological community quality. In addition, specific sample sites were needed to enable the evaluation of potential impacts separately. A reconnaissance of Salt Creek was also conducted and included attendance by NDEQ staff. Site selection incorporated the following factors:

- Complete characterization from the upper to the lower reach of Segment LP2-20000.
- Absence and presence of salinity.
- Absence and presence of channelization.
- Absence and presence of WWTP effluent.
- Seasonal variation, which includes evaluation of flow regimes and other environmental factors.

From 1994 through 1999, 10 instream chemical sampling sites, two WWTP effluent chemical sampling sites, and seven biological sampling sites were used to assess the chemical, physical, and biological aspects of Segment LP2-20000. Sample locations are shown in Figure 2-1.

Figure 2-1. Salt Creek Water Quality Studies Chemical and Biological Sampling Site.

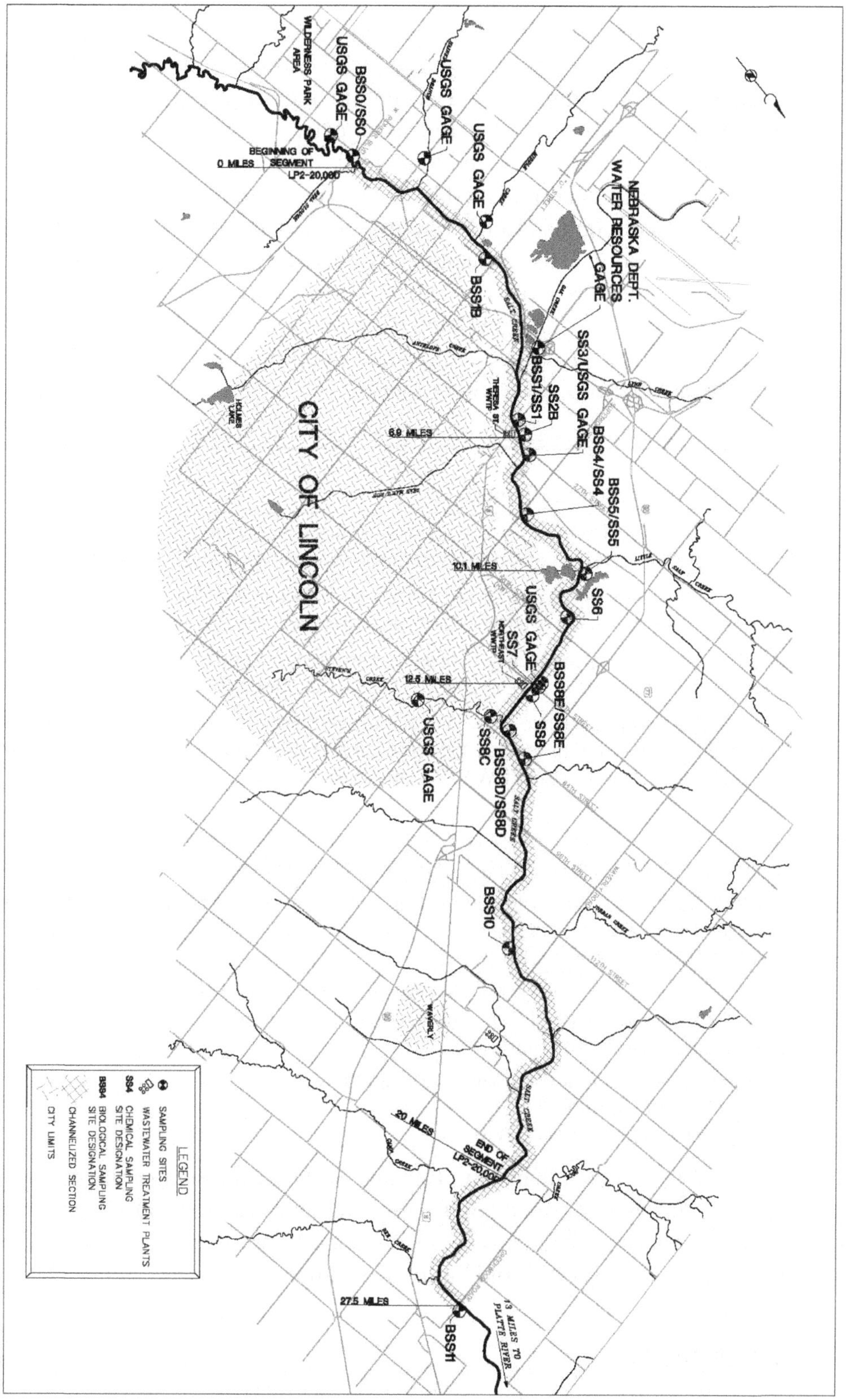

2.2 Biological and Chemical Assessments

From June 1994 through October 1999, six summer season and five winter season biological assessments were conducted on Segment LP2-20000 of Salt Creek. The biological assessments focused on the collection of fish and benthic macroinvertebrates, characterization of water quality, and habitat assessment. The technical objective was to characterize the condition of the biological community in Salt Creek and to identify the major limiting factors to the biological community. This was accomplished by evaluating both reference and impacted sites. A total of seven biological sampling stations were established along Segment LP2-20000 and sampled over the six year biological assessment study period. Each sampling station represents a different condition in Salt Creek, as shown in Table 2-1. In addition to biological sampling, commensurate chemical sampling was completed to characterize Salt Creek water quality at the time of the biological assessment. Additional chemical sampling was completed on a monthly to quarterly basis from June 1994 through October 1999.

Table 2-1. Salt Creek Chemical and Biological Sampling Sites and Purpose.

Sampling Station	Sample Site Description and Purpose
SS0/BSS0	RM 37.68. Wilderness Park (county rural area), upstream from all treatment plant discharges and not influenced by channelization or salinity. Measure non-impacted conditions. Control station for Salt Creek.
SS1/BSS01	RM 31.22. Urban area, upstream from all treatment plant discharges in the channelized and saline reach of Salt Creek. Measure impacts of channelization and salinity and possible impact of local industry.
SS4/BSS04	RM 29.04. Urban area approximately 10,000 feet downstream of the Theresa St. WWTP (RM 30.97) discharge in the channelized and saline reach of Salt Creek. Measure impacts of ammonia discharge.
SS6/BSS05	RM 27.01. Transitional area from urban to rural land use, approximately 21,000 feet downstream from the Theresa St. WWTP in the channelized and saline reach of Salt Creek. Measure rebound in biology below potential impacts of the Theresa St. WWTP ammonia discharge and reference condition above the Northeast WWTP.
SS8D/BSS08	RM 24.16. Rural area approximately 4,600 feet downstream of the Northeast WWTP (RM 25.03) discharge in the channelized and saline reach of Salt Creek. Measure impacts of ammonia discharge.
BSS10	RM 19.53. Rural area approximately 29,000 feet downstream from the Northeast WWTP discharge in the channelized and saline reach of Salt Creek. Measure rebound in biology below potential impacts of the Northeast WWTP ammonia discharge.
BSS11	RM 11.77. Rural area approximately 68,000 feet downstream of the Northeast WWTP discharge in the channelized and saline reach of Salt Creek. Measure of downstream condition after loss of instream ammonia with channelization and salinity.

Notes: RM = River mile from the confluence with the Platte River
SS0 = Designation for chemical sampling site.
BSS0 = Designation for biological sampling site.

As shown in Table 2-1, chemical sampling site SS0 and commensurate biological sampling site BSS0 are designated as the control, or reference site, for Salt Creek and are used for comparison to impacted sites. To measure the impact of ammonia from WWTP discharges, chemical sampling sites SS4 and SS8D and biological sampling stations BSS04 and BSS08 were used.

2.3 Collection Methods

Sample collection was specified in a Sampling and Analysis Plan prepared and submitted to the NDEQ. Bio-assessment procedures follow the technical guidance provided in the "Rapid Bio-assessment Protocols for Use in Streams and Rivers" (Plafkin et al., 1989); Revision to

Rapid Bio-assessment Protocols for use in Streams and Rivers: Periphyton, Benthic, Macroinvertebrates, and Fish (U.S. EPA, 1999); the NDEQ "Nebraska Stream Classification Study" (NDEQ, 1991); and "Draft Metric Revision for Nebraska Stream Classification Study" (NDEQ, 1995). A description of collection methods for chemical, physical, habitat, benthic, and fish communities is as follows:

- Chemical: Composite samples were collected at three equal-spaced locations along a transect.
- Physical/Habitat: Width, depth, and velocity were measured along three to five cross-sections at each station (Gordon et al. 1992). Three cross-sections were completed at each sampling station during each bio-assessment. Habitat quality was assessed visually and scored on standard field sheets by trained biologists using best professional judgment. Scores were recorded on Habitat Assessment Field Data Sheets developed by Plafkin et al. (1989) and modified by Barbour and Stribling (1994) and Environmental Monitoring and Assessment Program (U.S. EPA, 1999).
- Benthic and Fish Community: Benthic samples were collected from all available habitats with a 30-mesh (575 micron) D-frame dip net (Protocol III, U.S. EPA, 1996). Fish were collected by electro-fishing while wading and seining (Protocol III, U.S. EPA, 1996).

2.4 Chemical Condition Summary

Samples were collected for total ammonia and chloride analyses at stations BSS01, BSS04, and BSS08 during each bio-assessment, along with field parameters. Approximately 45 sets of chemical samples and field parameters were also collected at these stations during separate chemical sampling events from 1994 through 1999. Similar to hydraulic and habitat data, water chemistry provides information about the similarity of the living conditions for aquatic organisms among sites. In addition, the pH and temperature data document the range of exposure conditions that may have affected the relative proportions of ionized and un-ionized ammonia in Salt Creek. Total ammonia concentrations were measured on multiple occasions at levels that exceed the ammonia water quality criteria applicable to the Warmwater Habitat A classification.

Based on data from 1994 through 1999, ranges in total ammonia and chloride concentrations are as follows:

- Total Ammonia:
  - 0.1 mg/L to 1.0 mg/L in the reference reach at sample site SS0.
  - 0.1 mg/L to 10.0 mg/L below the Theresa Street and Northeast WWTPs at samples sites SS4 and SS8D.
- Chloride:
  - Less than 100 mg/L in the reference reach at sample site SS0.
  - 150 mg/L to 2,500 mg/L below the Theresa Street WWTP at samples site SS4 and 200 mg/L to 2,450 mg/L below the Northeast WWTP at sample site SS8D.

2.5 Physical and Habitat Condition Summary

As part of the biological assessment process, physical characteristics were measured including habitat quality, width to depth ratio, depth versus velocity, and substrate type. The hydraulic measurements, habitat scores, and chemical parameters were evaluated to determine the similarity of the physical and chemical conditions at the control station (BSS01) and the potentially impacted stations (BSS04 and BSS08). Differences in these variables among stations may cause changes in the biological community that would confound the interpretation of effects caused by ammonia toxicity. Conversely, the similarity of these variables among stations strengthens the assessment of ammonia impacts if ammonia is the only variable that changes significantly among the stations.

Median habitat quality scores ranged from 82 (BSS01) to 106.5 (BSS08), with a maximum possible value of 200. Habitat quality was judged to be poor because of the predominance of uniform depth and velocity; homogenous, unstable sand substrate; and the lack of pools and cover. Seventy percent of the substrate at Stations BSS01, BSS04, and BSS08 is sand, and hardpan clay makes up an increasing proportion of the substrate in downstream progression because of the scouring that occurs from high velocities.

A summary of Salt Creek physical and habitat characteristics is shown in Figure 2-2.

Figure 2-2. Physical Characteristics of Salt Creek Commensurate with Biological Assessments.

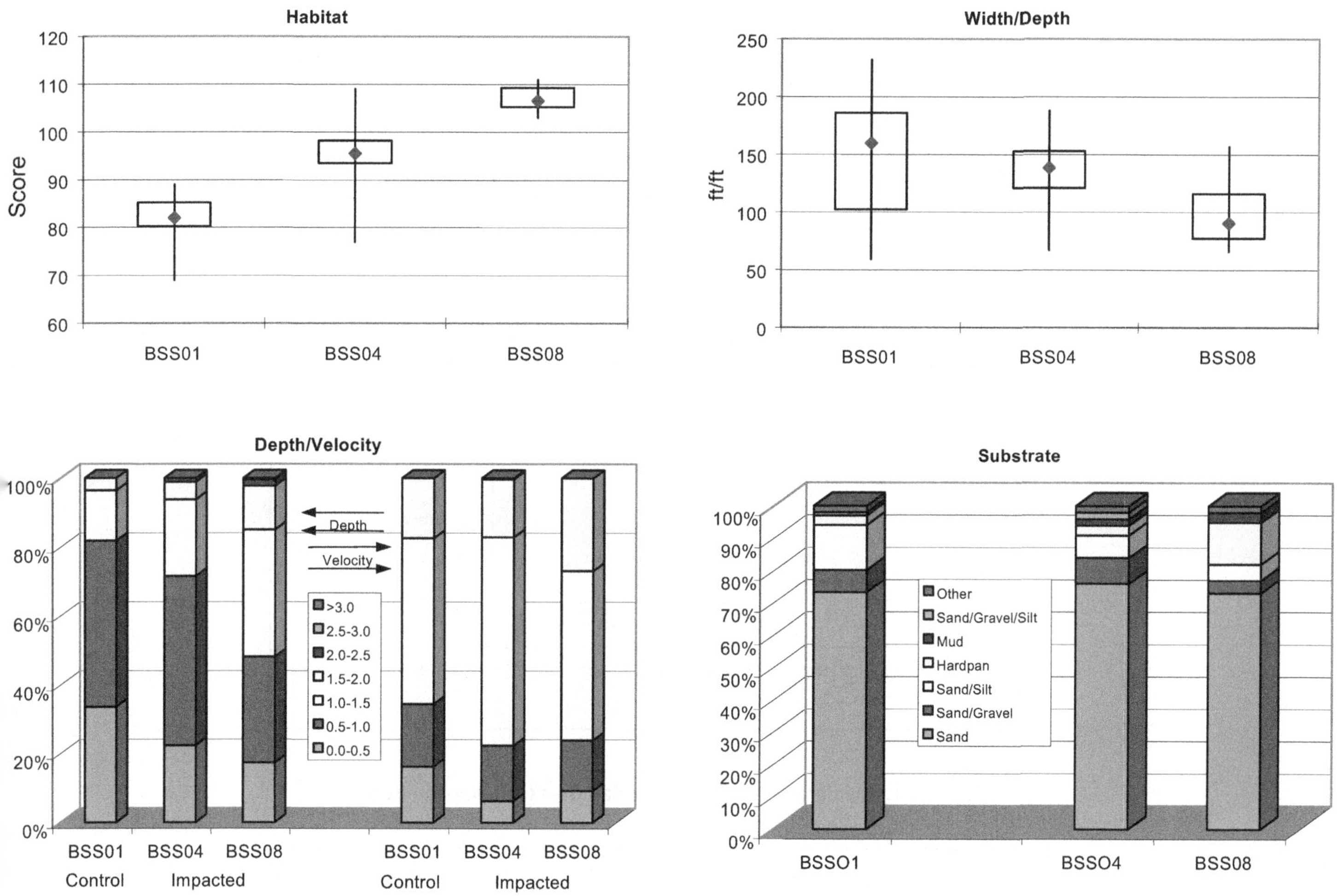

2.6 Biological Condition Summary

The Index of Biotic Integrity (IBI) is a multimetric score that combines various measures of fish species richness and composition, number and abundance of indicator species, trophic organization and function, reproductive behavior, fish abundance, and condition of individual fish. These multimetric scores were extensively as a tool for this project, as they indicate the overall health of the stream system. Box and whisker plots of the fish community condition, or IBI scores for the fish collections relative to habitat quality are shown in Figure 2-3. Based on the median scores, the fish community condition was poor at Site BSS0 and fair at the remaining sites. The IBI scores were considerably less than predicted by habitat quality at BSS0 and greater than predicted by habitat quality at Stations BSS01 through BSS11.

IBI scores for the impact and reference sites are presented in Figure 2-4. Median scores for reference sites and impact sites did not differ by more than three points for all the sampling events (summer and winter). The interquartile ranges between the reference and impacted stations broadly overlap, further suggesting a statistically similar condition of the fish communities between the two categories of sites.

Figure 2-3. IBI and Habitat Score Interquartile Ranges for Salt Creek, 7 Sampling Events.

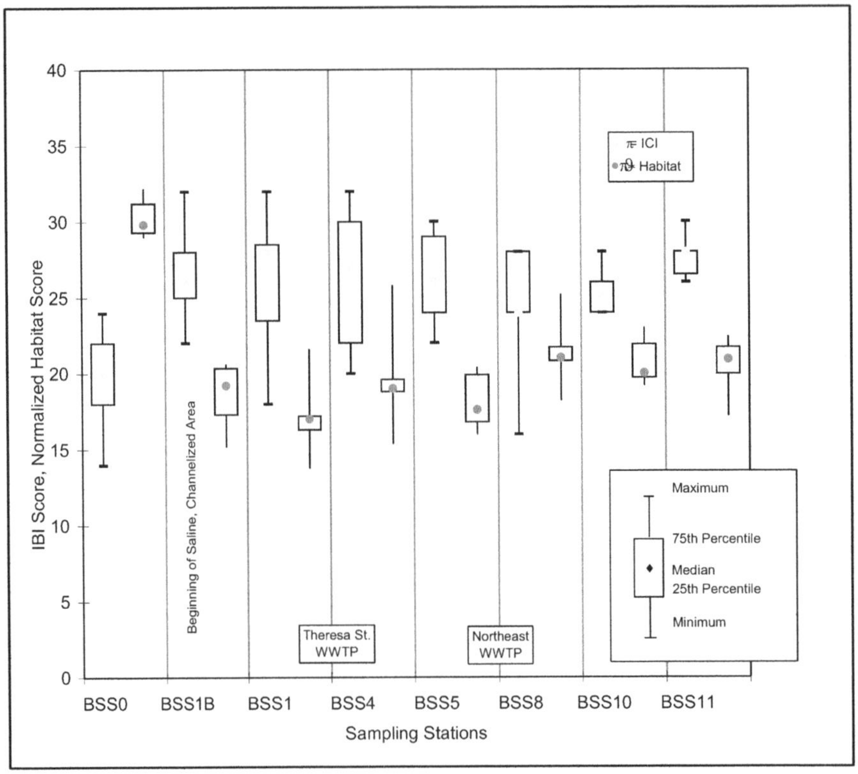

Figure 2-4. IBI Interquartile Ranges for Salt Creek, 7 Sampling Events.

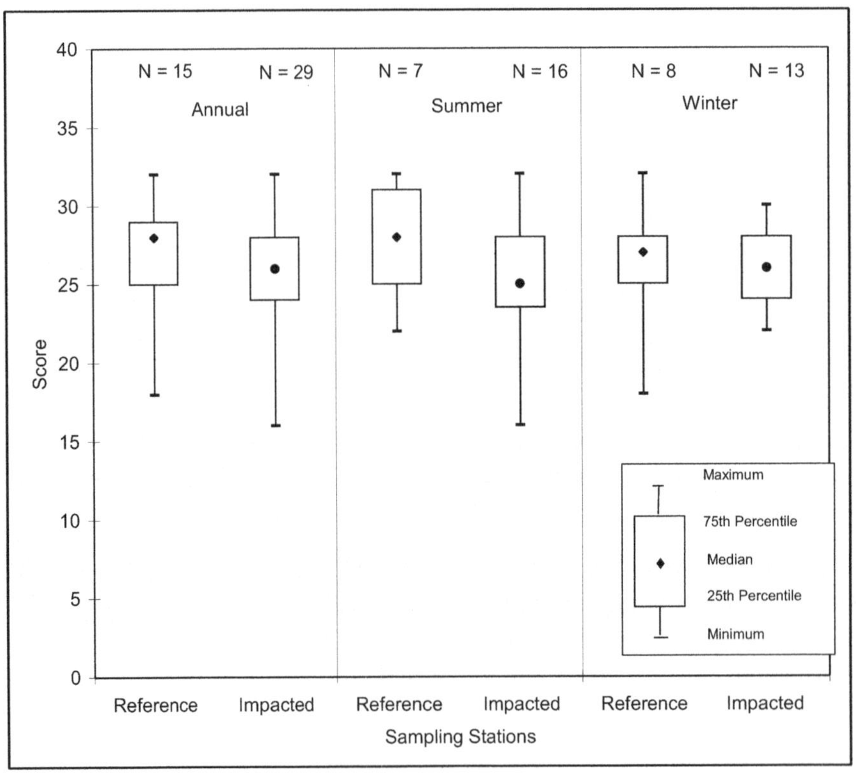

The relationship of benthic macroinvertebrate condition, or Invertebrate Community Index (ICI) scores for benthic macroinvertebrates and habitat quality is presented in Figure 2-5. The median ICI value is highest at site BSS0 (26) and decreases to 18 to 22 within the channelized reach. These values are 45-65% of the maximum score of 40 indicating that biological quality is fair to poor. The ICI interquartile ranges are shown in Figure 2-6.

Figure 2-5. ICI and Habitat Score Interquartile Ranges for Salt Creek, 7 Sampling Events.

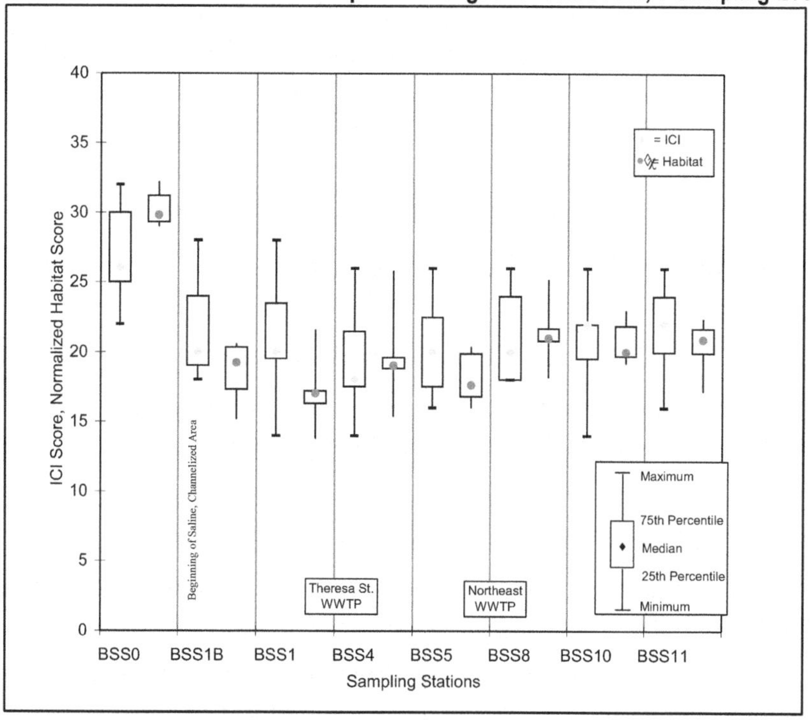

Figure 2-6. ICI Interquartile Ranges for Salt Creek, 7 Sampling Events.

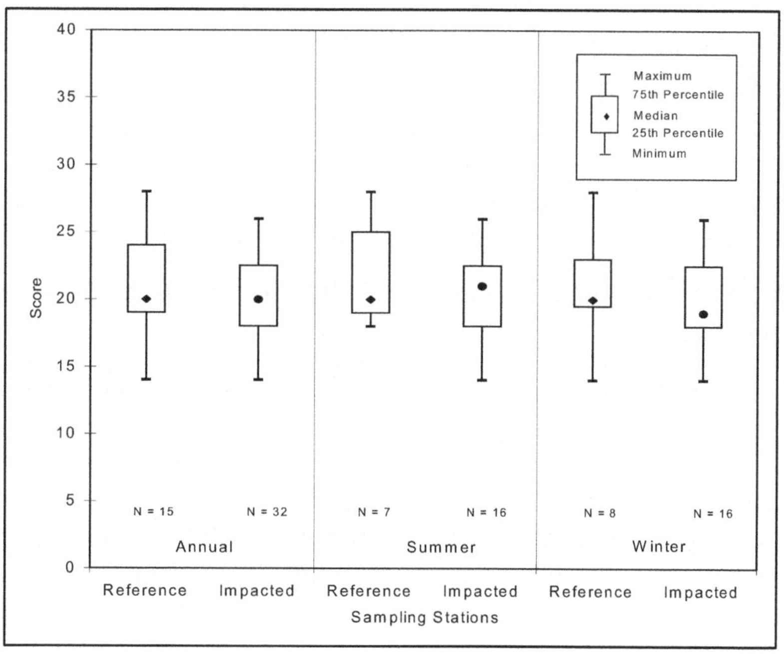

2.7 Summary of Key Findings from the Salt Creek Water Quality Studies

Results of the biological assessments, including characterization of Salt Creek water quality and habitat, indicated that there is no relationship between the fish community and benthic macroinvertebrate metrics and ammonia in Salt Creek. In addition, habitat quality was judged to be poor throughout the reach of Segment LP2-20000 from sample Station BSS1B, in the channelized and saline reach of Salt Creek above the WWTPs, through Station BSS11, the most downstream station, because of the predominance of uniform depth and velocity; homogenous, unstable sand substrate; the lack of pools and cover; and elevated salinity concentrations.

In summary, the key findings of the Salt Creek studies were:

- The Dakota sandstone formation is the predominant factor affecting water quality within the Segment LP2-20000 drainage through the addition of chloride.
- Flood control channelization has fundamentally altered the habitat in Salt Creek.
- Based on habitat quality, the Index of Biotic Integrity (IBI) index scores were less than predicted at the reference sites and greater than predicted at the ammonia-impacted sites.
- Channelization, chloride, temperature, and reservoir stocking are key determinant factors controlling biological communities in Salt Creek.
- Ammonia and several other constituents are not determinant factors for Salt Creek biological communities.
- The aquatic life use in Segment LP2-20000 is limited by naturally occurring chloride concentrations, and by human caused hydrologic modifications.
- The Class A Warmwater aquatic life use is not presently existing and cannot be attained in Segment LP2-20000 of Salt Creek.
- A site-specific application of water quality data is appropriate to develop site-specific Water Quality Standards.
- There is an inconsistency or disconnect between instream biological assessment data and national chemical specific data and criteria for ammonia. This disconnect is likely caused by many factors including the mitigative effects of salinity on ammonia toxicity, varying exposure, genetic adaptation, and the tolerance of the indigenous community.

2.8 Integration of Biological Assessment Results Into Ammonia Criteria Development

Throughout the period of conducting biological assessments on Salt Creek, the City evaluated ways to use the biological assessment results to derive, or to integrate into, the development of site-specific ammonia criteria for Salt Creek. This process included a collaborative stakeholder effort with the NDEQ Surface Water Section, Water Quality Division, and the U.S. EPA, Region VII, Water Quality Management Branch. During discussions and meetings with NDEQ and U.S. EPA, concerns arose over the technical implications of using biological data to develop numeric water quality criteria and the ability to integrate biological data under current NDEQ policy and regulation.

Under NDEQ regulations, specifically Title 117 - *Nebraska Surface Water Quality Standards*, procedures are identified for developing site-specific water quality criteria. These criteria follow standard U.S. EPA methods, such as the Recalculation Procedure, Water Effects Ratio, and Resident Species Approach. In addition to these standard methods, Title 117 regulations (Section 003.02A(g)(b)(4)) also provide for the use of "other scientifically defensible procedures such as relevant aquatic field studies, laboratory tests, or available scientific literature." Nationally, U.S. EPA recognizes the use of biological assessment data and information to develop biological criteria for regulatory purposes, specifically U.S. EPA 1996 *Biological Criteria for Streams and Small Rivers* guidance document (U.S. EPA, 1996).

2.8.1 Recommendations

The biological assessment results, including physical and chemical characterization, showed a disconnect between the historical levels of ammonia in Salt Creek and the existing biological communities. This disconnect did not support the need for the level of protection that would be provided by reducing ammonia effluent limits to the levels proposed in the draft NPDES permits. With concerns over using only biological assessment data to develop site-specific ammonia numeric water quality criteria, and recommendations from NDEQ, the City identified alternative methods for evaluating biological responses to ammonia concentrations in Salt Creek that more closely follow standard toxicological testing procedures, such as Whole Effluent Toxicity testing.

In discussions with NDEQ and U.S. EPA Region VII, the City proposed an *in situ* testing method to measure the sublethal chronic effects of the ambient Salt Creek ammonia concentrations on fish in Salt Creek. With this approach the historical biological assessment results could be used as corroborative evidence to support *in situ* study and further strengthen a biologically based approach to develop site-specific ammonia water quality criteria for Salt Creek.

CHAPTER 3.0

WATER ENVIRONMENT RESEARCH FOUNDATION PEER REVIEW PROCESS

WERF is a not-for-profit 501(c)(3) organization that was founded in order to fill the need for balanced water quality research addressing current wastewater research needs and forecasting future directions. WERF seeks cost-effective, publicly acceptable, environmentally sound solutions to water pollution control problems. In addition to providing peer review for WERF funded research, WERF also offers peer review services for subscribers who embark upon projects that have a research or demonstration aspect. These services are offered so that others may benefit from the results of the projects that are disseminated as WERF publications. Subscribers request peer review because they are seeking high quality, national, independent analyses of their research projects. WERF peer review enhances results for both the decision makers and the public by ensuring the scientific work has been done reliably and responsibly. This promotes acceptance of the resulting decisions or future projects, builds confidence in the underlying assumptions on which recommendations of a report are based, and can add to an organization's credibility by evaluating the scientific validity of its research results before releasing them to the public.

3.1 City of Lincoln Peer Review Request

The City recognized that in order to develop site-specific criteria for ammonia the technical approach needed to be scientifically sound and defensible and fit in to the current regulatory framework. The City proposed the SCWQS project for WERF peer review, expecting that such a review would make the approach to developing a site-specific ammonia criteria robust and defensible and would provide an independent perspective.

The important nature of the SCWQS, and the proposed *in situ* toxicity test, plus the usefulness of information from the study to other subscribers, made it a compelling project for WERF peer review. At the same time, the existence of regulatory jurisdictional matters beyond the scope of WERF's review required close oversight of this project for objectivity, independence, and strict adherence to the scope. For these reasons, two Research Council liaison/members sat on the Peer Review Panel as Co-Chairs:

- Robert Berger – Manager Regulatory Planning and Analysis Manager, East Bay Municipal Utility District, Oakland, California
- Norm LeBlanc – Chief of Technical Services, Hampton Roads Sanitation District, Virginia Beach, Virginia

The additional members of the Peer Review Panel were:

- Scott Belanger, Ph.D. – Principal Scientist, Procter & Gamble, Cincinnati, Ohio

- Jerry Diamond, Ph.D. – Principal Aquatic Ecotoxicologist, Tetra Tech, Inc., Baltimore, Maryland
- James Karr, Ph.D. – Professor of Fisheries and Zoology, University of Washington, Seattle, Washington
- Thomas Sprehe – Chief of the Environmental Engineering Division, KCI, Hunt Valley, Maryland

3.2 Peer Review Panel Charge

At the initial January 1999 site visit, the Peer Review Panel and the City reviewed and agreed upon the following charge to the Panel:

Review and Assess:

- Whether the methodologies and approach for conducting biological assessments in Salt Creek are scientifically defensible for assessing:
    - Effects (acute, chronic sublethal, and other) of human activities on Salt Creek aquatic life.
    - Effects (acute, chronic sublethal, and other) of WWTP effluent on Salt Creek aquatic life.
    - Effects (acute, chronic sublethal, and other) of ammonia on Salt Creek aquatic life.
- Whether the available Salt Creek biological, physical, chemical, and toxicological data are scientifically defensible for assessing:
    - Effects (acute, chronic sublethal, and other) of human activities on Salt Creek aquatic life.
    - Effects (acute, chronic sublethal, and other) of WWTP effluent on Salt Creek aquatic life.
    - Effects (acute, chronic sublethal, and other) of ammonia on Salt Creek aquatic life.
- Whether the methodologies for assessing acute, chronic sublethal, and other effects using biological assessment, *in situ* chemical or toxicity testing, and other data are scientifically defensible.
- Whether the use of biological assessment, *in situ* testing, and other Salt Creek data and information and methodologies for their integration are scientifically defensible to establish site-specific chronic water quality standard criteria for ammonia on Salt Creek.
- Whether quality assurance/quality control procedures are appropriate.

The Peer Review Panel attended two workshops held at the City of Lincoln, which were also attended by City personnel, City contractors, and representatives from NDEQ and U.S. EPA Region VII. In addition to the workshops, multiple conference calls and e-mail exchanges took place in a collective effort to address suggestions and comments from the WERF Peer Review Panel, NDEQ, and U.S. EPA.

3.3 WERF's General Appraisal of the Studies and Response to the Charge

In addressing the Peer Review Charge and the Salt Creek studies as a whole, the Peer Review Panel found that the studies and quality of the data collected were of sufficient technical and scientific validity. From this, the Peer Review Panel was able to state that the results of the studies, overall, are scientifically defensible for the purposes they were designed.

Recommendations and critiques were provided, specifically for the *in situ* toxicity testing procedures, which were incorporated. It was determined that the methodologies used by the City in assessing acute and sublethal toxic effects are routinely used by resource agencies, researchers in academia and industry, and others in aquatic resource management. The primary strength of the Salt Creek studies is the ability to use several different lines of evidence to evaluate effects, and it is possible to develop a weight-of-evidence evaluation that is scientifically defensible. The use of several tools to derive an ammonia criterion, as was done by the City, adds strength to the derivation of a criterion.

3.4 WERF Peer Review Recommendations

One of the key outcomes of the WERF Peer Review was the support to perform *in situ* testing to measure biological response directly in Salt Creek rather than depending on extrapolating laboratory results and comparing them to biological assessment data. The WERF Peer Review Panel recommendations for developing a technically supporting *in situ* study included collecting data and information with sufficient precision to meet the following objectives:

- Characterize the levels and variability of chemical parameters of concern that occur during the *in situ* exposure of the selected fish species.
- Compare chemical parameter results to those measured during historical biological assessments.
- Link ammonia levels to sublethal (chronic) responses of the selected fish species.
- Evaluate the effects of ammonia on two species of fish with known sensitivity to ammonia.
- Measure toxicity exposures at a sufficient number of locations along Salt Creek along an ammonia concentration gradient.
- Statistically distinguish chronic responses between control and test sites.
- Develop relationships between measured ammonia levels and response of exposed fish.
- Link fish response to identified human activities on Salt Creek.
- Calculate an IC20 (Inhibition Concentration that affects 20% of the fish population) for each fish species.

Based on the WERF Peer Review Panel recommendations, the City of Lincoln initiated *in situ* testing in Salt Creek to more accurately characterize the complex interactions of ammonia in Salt Creek waters and support the development of a site-specific chronic ammonia criterion.

CHAPTER 4.0

SALT CREEK *IN SITU* TOXICITY TESTING PROGRAM

To more accurately characterize the complex interactions of ammonia in Salt Creek, an *in situ* toxicity test was performed in which caged fish were continuously exposed for 15-day and 30-day periods at nine in-stream locations. The *in situ* program was developed by the City and their consulting team (Brown and Caldwell; EA Engineering, Science & Technology, Inc.; and CH2MHill), with technical input from the WERF Peer Review Panel. Technical and regulatory support for the development of the *in situ* study program was also provided by the NDEQ, Surface Water Section, Water Quality Division, and U.S. EPA Region VII, Water Quality Management Branch.

Although the final *in situ* program was carried out in the fall (October) of 1999, extensive preliminary research preceded the *in situ* test. Specifically, an experimental test protocol was developed for *in situ* testing with channel catfish and the test protocol was then refined in a three-phase study, as follows:

- Phase 1: Laboratory testing involving the design and development of the testing protocol, including selecting test species, designing test chambers, and developing test procedures.
- Phase 2: "Pilot" *in situ* study to implement and refine the methods developed in the laboratory, and to determine if the test fish would show significant growth during a four-week exposure period. From this study, modifications were determined and the *in situ* test procedure was adjusted accordingly.
- Phase 3: Laboratory and field studies to determine the suitability of fathead minnows and channel catfish for the proposed *in situ* toxicity testing program.

To document the proposed *in situ* testing methods development and full-scale application in Salt Creek, the City developed a comprehensive Work Plan, which was reviewed by NDEQ, U.S. EPA, and the WERF Peer Review Panel (City of Lincoln, 1998).

4.1 *In Situ* Toxicity Testing Methods Development

A two-phase study was conducted to develop a test protocol for *in situ* testing with channel catfish. The first phase was conducted in a laboratory and involved the design and development of the protocol, including selecting a test species, designing a test chamber, and developing test procedures. The second phase was a pilot *in situ* study to implement and refine methods developed in the laboratory, and to determine if the test fish would show significant growth during a two-week test period. From this study the needed modifications were determined and the test procedure was adjusted accordingly.

4.1.1 Method Development in the Laboratory

This phase of the study was conducted at the EA Engineering, Science and Technology Ecotoxicology Laboratory in Baltimore, Maryland. Many of the tasks were conducted concurrently in order to make adjustments in design and procedure, and to complete the study in a short time period. The factors considered critical to developing a successful study protocol for the *in situ* caged fish studies design were as follows.

4.1.1.1 Fish Species Selection

The selected fish species needed to be representative of the fish community in Salt Creek, sensitive to ammonia, able to tolerate ambient Salt Creek salinities, and behaviorally adaptable to being held in a cage or test chamber. The channel catfish (*Ictalurus punctatus*) met all these criteria and was chosen as one of the species well-suited for the in situ testing. In addition, channel catfish are listed as a "key" species in Salt Creek under NDEQ Title 117 regulations. Channel catfish are easily obtained from aquaculture facilities, are conditioned to feeding on commercially prepared feeds, and are not hypersensitive to handling stress. In addition, they are very gregarious, and the species was expected to perform well under the conditions in an enclosed test chamber. Juvenile fish were chosen for testing since they tolerate handling well (better than larvae), and they are large enough to feed on commercial fish food but small enough that they should achieve measurable growth over a relatively short period of time. The channel catfish has been used successfully for other NPDES compliance programs nationwide.

4.1.1.2 Exposure Period

Selecting an in-stream exposure period to minimize damage to the organisms from high flows during major storm events or during low flow periods but long enough to achieve measurable fish growth was important. To determine the time period to achieve measurable growth, channel catfish approximately two inches in length were obtained from a commercial supplier and acclimated to laboratory conditions. The fish were sorted to obtain fish of uniform size and placed in flow-through, two-gallon aquariums to evaluate growth rate. Initial lengths and wet and dry weight measurements were made on representative specimens. Ten aquaria were set up with 10 fish per tank, and ten aquaria were set up with 15 fish per tank, in order to investigate if growth rate would vary with density of fish. The fish were fed 2.5% of their body weight twice a day with a floating catfish chow provided by the aquaculture facility. Twice a week one tank of fish (10 and 15 fish) were harvested and the fish were measured to evaluate growth. Initially the test was set up with freshwater, but because of mortality in some test aquaria due to disease problems, the water was changed to five parts per thousand (ppt) salinity, which is within the range that is found in Salt Creek. This controlled the disease and 5 ppt water was used for all further laboratory testing. All testing was done with water temperature set at 20 degrees Celsius and a 16-hour light/8-hour dark light cycle. The results of the growth test indicated significant growth by a minimum of 7 days, and 14 days was selected as the final test duration in order to obtain more data to strengthen the test conclusion.

4.1.1.2.1 Test Chamber Initial chamber and procedure development was tested in the laboratory in a Living Stream unit, which has a re-circulating laminar flow. An *in situ* test chamber was needed that would provide adequate flow to maintain acceptable water quality and provide representative exposure to Salt Creek ambient conditions. The test chamber design also had to facilitate daily feeding, maintenance, and observation of test animals. This was accomplished by designing a test chamber using polyvinyl chloride (PVC) well screening with an end cap at one end and a threaded coupling and removable plug at the other end. An eye bolt was fastened to the threaded plug providing an attachment for mooring lines. The initial design utilized 4-inch diameter PVC with 0.020-inch slots. It was determined that flow was inadequate due to excessive fish waste build-up and water quality related mortality occurred in the 4-inch diameter chamber. A 16-inch length of 6-inch diameter PVC well screen with a larger slot size (0.30 inch) was selected for the final design (Figure 4-1). This chamber had approximately a 2-gallon capacity comparable to the aquaria used for the initial growth test. The larger slot size allowed better flow through the chamber but still retained the feed.

Figure 4-1. In Situ Testing Fish Chamber Diagram.

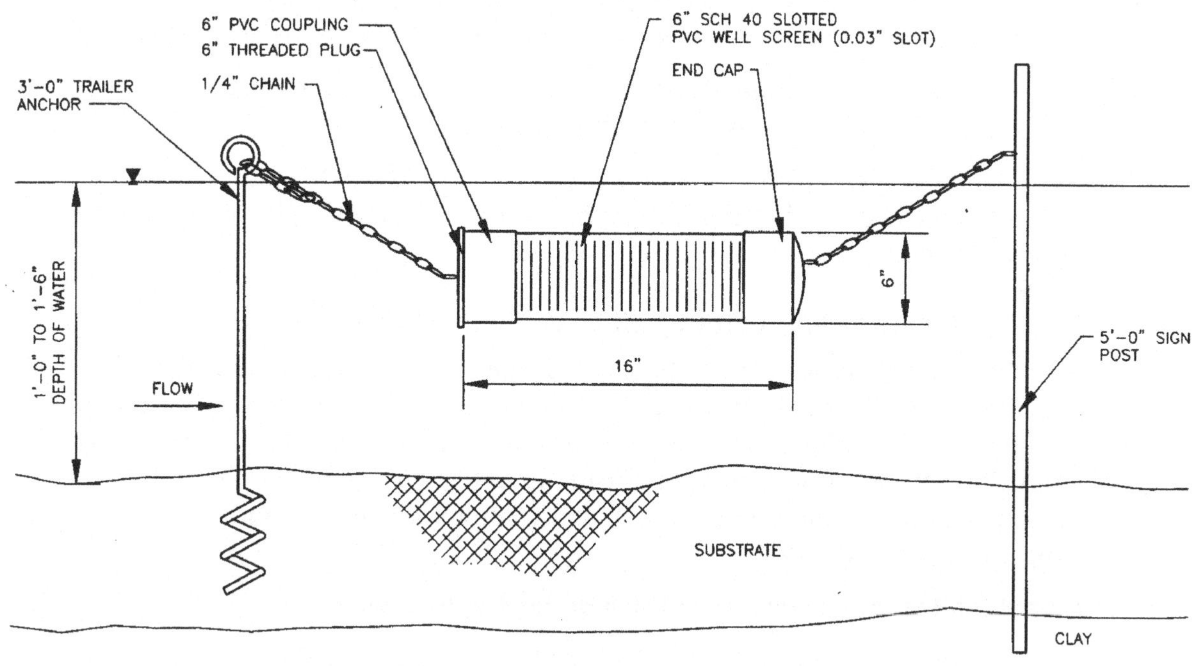

4.1.1.2.2 Field Test and Feeding Regime Development The prototype test chamber was deployed in a stream near the Ecotoxicology Laboratory in Sparks, Maryland to evaluate the chamber design and maintenance procedure in a "real world" situation where daily temperature varied, silt and sand were present (especially after storm events), and flow rate was variable both in time and position or placement of the chamber in the stream. Daily maintenance included cleaning the slots, removing fish waste and silt from the inside of the chamber with a turkey baster, observing behavior, and feeding twice a day. Three 6-inch diameter chambers were then deployed in the stream and the fish were fed sinking trout chow at the established rate. After one week, one chamber was removed from the stream and fish measured. The comparison with initial wet and dry weights indicated growth had occurred.

4.1.1.3 Supplemental QA/QC Testing Extensive efforts were conducted to ensure that data quality would be acceptable in a research/regulatory program of this type. These efforts included:

- Salinity tolerance testing of 30- to 45-day old fathead minnows and channel catfish.
- Salinity intolerance testing of less than 24-hour old fathead minnow larvae.
- Test organism age study to quantify the sensitivity of less than 24-hour old fathead minnows versus 30- to 45-day old fathead minnows to ammonia.
- Examination of the quality of test organisms using reference toxicant testing.
- Parallel chronic WET testing (during *in situ* program).
- An *in situ* salinity tolerance study using a series of control stations in several salinity zones (e.g., 1, 2, and 4 ppt, located upstream of the Theresa Street WWTP).
- Dye testing of exposure chambers to quantify the turnover volumes to ensure there was sufficient flow through the chambers.
- Numerous field quality assurance/quality control (QA/QC) issues including randomizing cage locations, cage maintenance staff, and staff responsibilities within each team.

4.1.2 Method Development in Salt Creek – Pilot Test

The second phase of the methods development study was conducted in Salt Creek. The 6-inch diameter test chamber prototype was deployed in Salt Creek to test the *in situ* methods. The pilot study was successful and demonstrated that the *in situ* method is a reliable method for measuring in-stream toxicity. Site selection, chamber development, acclimation, test initiation, daily monitoring, and test termination followed the specifications described below.

4.1.2.1 Pilot Test Site Selection and Chamber Deployment

Fifteen test chambers were constructed by City personnel based on the final chamber design (Figure 4-1). Chambers were located at three distinct locations in Salt Creek to ensure enough data for analysis despite vandalism and the potential loss due to floating debris. Selection of test location sites was dictated by the stream conditions, including adequate depth, visual separation of test sites, and stream attributes to maintain depth. All sites were located in the channelized, saline, non-wastewater impacted segment. Chambers were deployed in Salt Creek with five replicate chambers per test location. Each chamber was independently suspended at mid-depth in 1-1.5 ft of water. Chambers were anchored using a sign post and a trailer anchor driven to a depth of competent substrate (clay). Early deployment of the chambers was done so they could acclimate to Salt Creek water and to remove any residual PVC contaminants.

4.1.2.2 Pilot Test Site Results

The pilot test in Salt Creek showed that the growth of channel catfish is inversely related to salinity and the relationship is predictable, showing a cause and effect relationship with salinity. The pilot showed that channel catfish were well-adapted to the test chambers and grew by 20-35% (dry weight) over the 14-day pilot test period. This growth was strong evidence of the adequacy of the chambers and the sufficiency of the feeding regime. Build-up of wastes, food, and sediment was minimal indicating good water flow through the chambers. Maintenance procedures were routine and predictable and the chambers facilitated regular observation of the test animals without stress to the fish.

There was limited variability in growth during the Salt Creek pilot test for this method. The coefficients of variation (CVs) for the results from the three Salt Creek test sites varied from 14-17%, whereas the CVs for the methods development tests conducted in the laboratory ranged from 14-38%. In contrast, the CV in the control treatment for enforceable effluent biomonitoring (WET testing) can be as great as 40% and the quality of the test results is acceptable for measuring effluent toxicity.

Based on the results of the pilot test, it was determined that the methods developed would apply to a full-scale, 15-day and 30-day duration, in situ test in Salt Creek producing accurate and precise measurement of the sub-lethal chronic toxicity to the test species.

4.2 Salt Creek Full-Scale *In Situ* Toxicity Testing

Full-scale *in situ* toxicity testing occurred in Salt Creek during the month of October 1999. The location of *in situ* testing fish chambers, or cages, was based on target total ammonia concentrations (and salinity) in Salt Creek. The target ammonia concentrations needed to include a chronic effect level for a 30-day exposure period to fathead minnows and channel catfish, so that the test results could be used to develop a 30-day EC20 value. Based on the 1998 Addendum to the U.S. EPA (1984) Ammonia Criteria Document, the chronic effect level is approximately 3.09 mg/L total ammonia for the fathead minnow, and 8.84 mg/L for the channel catfish. To develop an EC20 value, or an alternative endpoint that might better predict chronic threshold effect concentrations to resident communities, *in situ* test cages were placed in Salt Creek at locations covering a range of in-stream total ammonia concentrations. The locations of the test cages were chosen to bracket the range of EC20 concentrations that have been reported for total ammonia ranging from less than 3 mg /L to about10 mg /L total ammonia.

Five locations downstream of the Northeast WWTP were used (Stations E, F, G, H, and I), in addition to one ammonia "control" station immediately upstream of the Northeast WWTP (Station D). To evaluate the effect of salinity, three additional upstream salinity "controls" were located above the Theresa Street WWTP (Stations A, B, and C) to reflect the potential range of salinities at the downstream stations. The test chamber locations below the Northeast WWTP were selected for three primary reasons:

1. In-stream ammonia concentrations are generally higher than below the Theresa Street WWTP, as required to meet the necessary chronic toxicity exposure range.
2. Salt Creek is deeper below the Northeast WWTP, which is necessary to properly locate the *in situ* testing chambers.

The chemistry and WET test data for the Northeast WWTP shows less variability, which enhances the quality of the research.

This experimental design resulted in a total of nine sites, as shown in Figure 4-2.

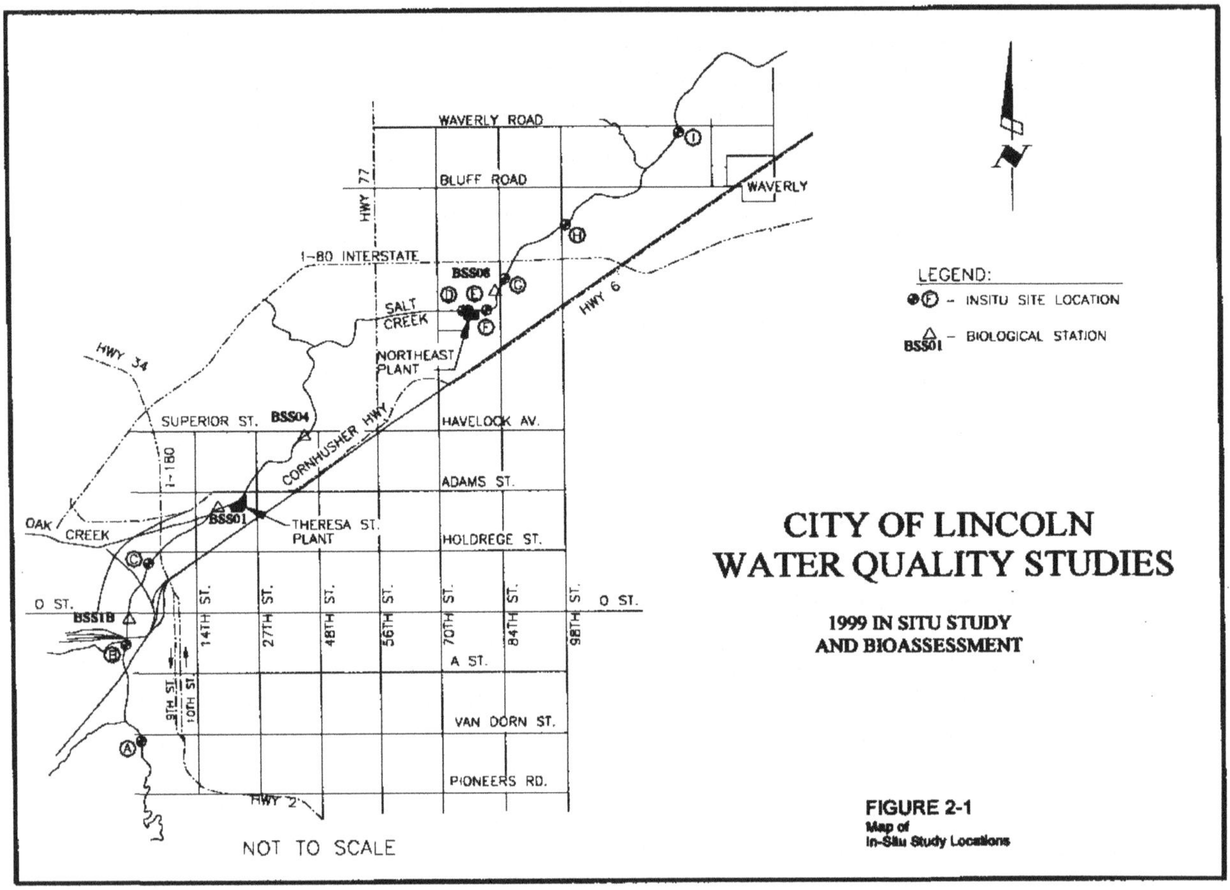

Figure 4-2. Location of *In Situ* Fish Chamber Stations in Salt Creek.

A description of the purpose of each *in situ* site location is provided in Table 4-1.

Table 4-1. Description of Station Locations for the Salt Creek *In Situ* Study.

Station Name	Description of *In Situ* Sites
A	Reference site upstream of Haines Branch confluence. Conductivity <1000 μS/cm and ammonia concentrations at background levels.
B	Reference site, conductivity same range as downstream sites (4,000 – 5,000 μS/cm). Ammonia concentrations at background levels.
C	Reference site, conductivity 6,000 – 7,000 μS/cm, ammonia concentrations at background levels. Upstream of the Theresa Street WWTP.
D	Immediately upstream of the Northeast WWTP outfall, control site for effects from Theresa St. WWTP outfall, Goodyear ditch, etc. Had elevated ammonia concentrations during study (5–7 mg/L)
E	The site with highest ammonia concentrations. Approximately 70 meters downstream of the Northeast WWTP outfall.
F	Approximately 600 m downstream of the Northeast WWTP outfall, upstream confluence with Stevens Creek
G	84th Street Bridge approximately 1,500 meters downstream of the Northeast outfall
H	98th Street Bridge approximately 3,645 m downstream of the Northeast WWTP outfall. No fathead minnows at this site.
I	Approximately 8,000 meters downstream from the Northeast WWTP outfall located at Waverly Road.
	Description of Supporting Sampling Sites
Theresa Street WWTP Outfall	WWTP between Stations C and D
Northeast WWTP Outfall	WWTP just downstream of Station D
USGS 27th St. (Site 3)	USGS gauging station located near 27th Street Bridge. Velocity/flow transects and chemical sampling were also conducted at this site for studying relationships with the Theresa St. WWTP effluent.
USGS 70th St.	USGS gauging station located at 70th Street Bridge

4.2.1 *In Situ* Testing Sample Size

Sample size requirements for *in situ* testing were defined to detect differences between two populations with equal variances and account for the site-specific application to Salt Creek. The sample size required at each *in situ* station was based on data collected from the pilot growth study conducted in Salt Creek. The replicate sample size was 8 test chambers per site (per species) with 15 fish (fathead minnows or channel catfish) in each chamber. The design allows the detection of a 20% difference in growth among stations (during a 30-day test) with a 90% confidence.

Due to the possibility of severe weather and high flows that could threaten the entire field program, a decision was made to deploy a total of 32 chambers per station (16 chambers for fathead minnows and 16 for channel catfish). This approach allowed for half of the chambers (8) for each species to be "harvested" and yield a valid test anytime after 15 days, leaving the remaining 16 chambers in stream for the full 30 days. A general diagram of a fish chamber layout is provided in Figure 4-3, including the chambers removed after the initial 15 days of the study.

Figure 4-3. In Situ Toxicity Testing Fish Chamber Configuration.

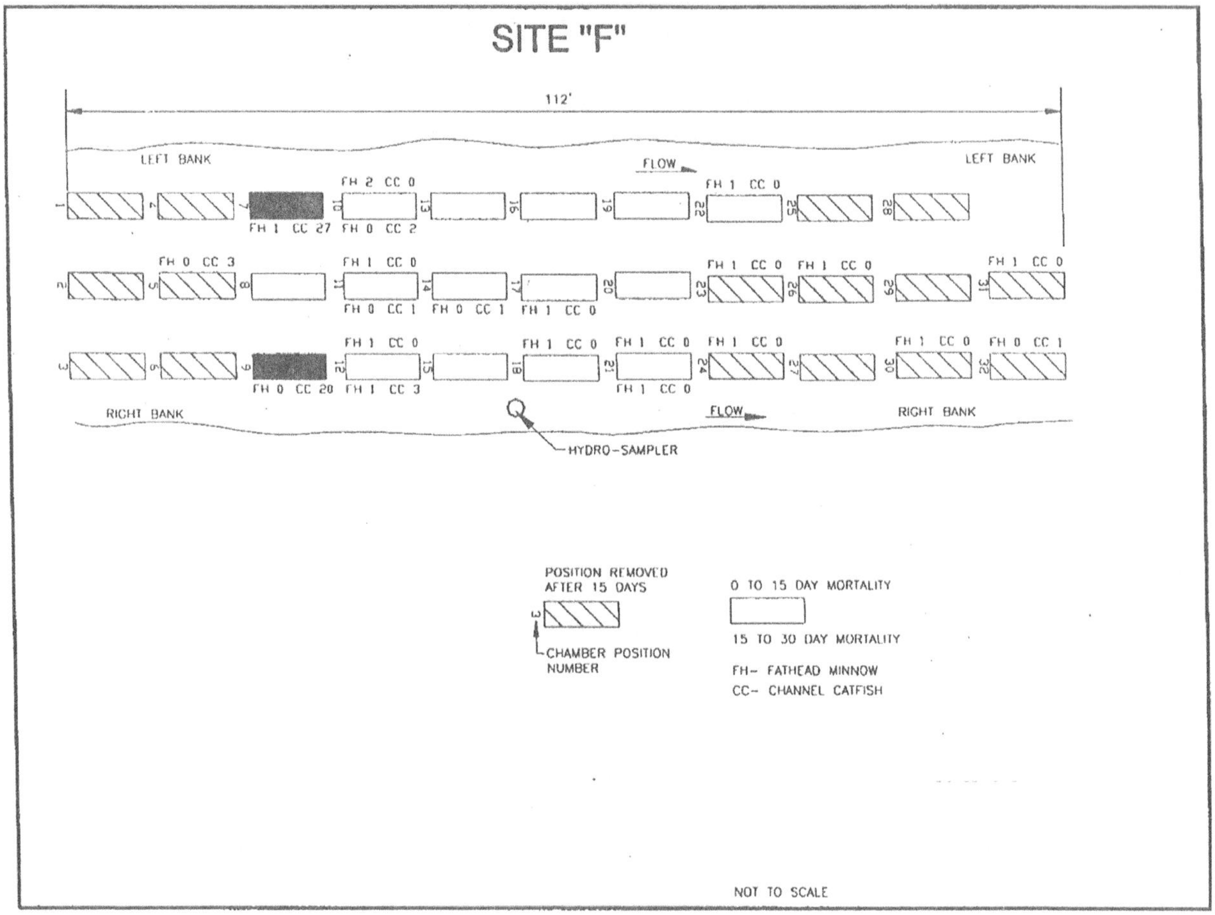

4.2.2 Fish Acclimation

Prior to receipt of the catfish, a three hundred gallon vinyl tank was set up near Salt Creek, upstream of the WWTPs, and used for acclimation of the catfish to the water of Salt Creek. Three hundred 2-inch channel catfish were received from the supplier and, upon receipt, the catfish shipping container was opened and air lines were placed in the water to provide air and remove waste. Channel catfish were then observed for one hour for signs of stress and shipping mortality. The plastic bags holding the channel catfish and shipping water were then placed in the tank to allow water temperature to slowly equilibrate. Throughout this process water temperature was monitored and the fish were released into the tank when there was less than two degrees of temperature difference.

Salt Creek water was pumped to the tank via a submersible pump in the creek. Throughout the acclimation period, water was drained from the tank and renewed with creek water. During this process, water quality was monitored, particularly conductivity as an indirect measure of salinity. The acclimation was considered complete when the conductivity in the tank was equal to that of Salt Creek. The channel catfish were observed for signs of stress, disease, and mortality, as well as feeding behavior during the acclimation period. No mortality or signs of stress were noted. After 48 hours the channel catfish were placed in the test chambers.

4.2.3 *In Situ* Test Initiation

A total of 20 channel catfish and 20 fathead minnows were randomly selected from the large holding tank prior to placement in the test chambers to establish initial fish lengths and weights prior to initiation of *in situ* testing. During the selection process, fish that were obviously smaller or larger than average were returned to the tank. The selected fish were taken to the City's WWTP laboratory where they were weighed and measured. All weight and length measurements were recorded on standard laboratory sheets with weights measured on a Mettler H10T analytical balance. Total length of each fish was measured using calipers. Wet weights were measured using pre-dried and pre-weighed aluminum tins. Fish were dried for 24-hours in an oven at 100° Celsius, after which the fish were reweighed to obtain dry weights.

4.2.4 Daily Monitoring Activities

Maintenance of chambers and feeding of fish was performed two times a day (7:30 a.m. and at 4 p.m.) for the duration of the study. Chamber maintenance, feeding, and water quality measurements were completed in the same sequential order on each visit. Water quality parameters, including dissolved oxygen, temperature, pH, conductivity, depth, and velocity were collected each day at the upstream end of the sampling station. Daily maintenance of the chamber included scrubbing the inside and outside to remove algae and sediments. As needed, excess sediment in the chamber was removed. Channel catfish and fathead minnows were fed 2.5% of their initial wet body weight with sinking salmon starter chow. The fish were then observed to evaluate feeding behavior and indications of stress or mortality.

4.2.4.1 Chemical Monitoring

Extensive chemical monitoring was conducted both in Salt Creek and within the Northeast and Theresa Street WWTPs' final effluent during *in situ* testing in Salt Creek. Parameters monitored in Salt Creek included: total ammonia, chloride, pH, conductivity, temperature, flow, velocity, air temperature, and precipitation. Details on the locations and frequency of these measurements are presented in Table 4-2.

Table 4-2. Summary of *In Situ* Study Primary Monitoring Parameters.

Parameters	Location	Collection Frequency
Flow, Velocity	*in situ* stations, WWTP outfalls, USGS gauging stations	Continuous metering at WWTP outfalls and USGS gauging stations; daily transects for *in situ* stations D, F, G, and I.
Air Temperature, Precipitation	*in situ* stations, WWTP outfalls	Continuous metering at University of Nebraska High Plains Climate Center weather station sites.
pH, Conductivity, Temperature, Dissolved Oxygen	*in situ* stations	Continuous metering (except for dissolved oxygen) with HydroLab® units at all *in situ* stations; portable meter spot checks twice per day at *in situ* stations A, B, and C and once per day at *in situ* stations D–I.
Total Ammonia, Chlorides	*in situ* stations, WWTP outfalls	Every 4 hours for *in situ* stations D–I and twice daily at *in situ* stations A, B, and C.

In addition, during *in situ* testing, limited pesticide and herbicide sampling was conducted to evaluate other significant factors that could affect the *in situ* test results. Pesticide and herbicide samples were collected at three times during the 30-day test at three locations: 1) upstream of the Theresa Street WWTP at *in situ* Station C; 2) just above the Northeast WWTP at *in situ* Station D; and 3) below the Northeast WWTP at *in situ* Station I.

4.2.5 Fish Mortality Evaluation

Each fish chamber, at each *in situ* station, was evaluated twice per day during feeding and maintenance activities throughout the 30-day exposure period, and mortalities in each chamber were recorded. In addition, to help ensure that organism exposure was not influenced by chamber location relative to the other *in situ* exposure chambers, the position of the chambers within the array at each site was changed daily. Chamber positions within each array were numbered (1 through 32) for each of the *in situ* study stations. Chambers located in an odd position were rotated daily downstream to the next odd position and likewise with chambers occupying even numbered positions. Thus, a chamber at position 1 at the initiation of the study would have occupied every odd position in the array at the end of the 15-day test.

4.2.5.1 15-Day Mortality Results

For the first 15 days, mortality upstream and downstream of the Northeast WWTP was low. The mean percent mortality for fathead minnows was 3.2% (Table 4-3), and the mean value for channel catfish was 0.5% (n = 9). Further, the limited mortality that did occur during this period appeared to be unrelated to ambient ammonia concentrations, since the *in situ* sites with the highest mortality for fathead minnows (Stations A and B) are located above both WWTPs. For channel catfish, the mortality was very low at all sites.

4.2.5.2 30-Day Mortality Results

For the full 30-day *in situ* test, mean mortality for fathead minnows averaged 6.8%, with the highest mortalities again at upstream Stations A and B (above the WWTPs), and at downstream Stations E and F (Table 4-4). The channel catfish data were more difficult to interpret since there was limited mortality at all stations except Stations E and F, which are immediately downstream of the Northeast WWTP. At Station F, mortality was 42%, with almost all mortality occurring in only three of the eight fish chambers. At Station E, mortality was 44.8%, with almost all mortality occurring in two of the eight fish chambers. If these fish chambers are removed from the percent mortality calculation, the mortality rate at Station E is reduced from 42% to 14.3%, and 44.8% to 18.3% at Station F. Mortality rates for channel catfish are shown in Table 4-4. The locations of the high mortality fish chambers, for both species, are shown in Figures 4-4 and 4-5.

Table 4-3. 15-Day Fish Mortality.

	Fathead Minnow			Channel Catfish		
Station	Percent Mortality	Standard Error	# Chambers	Percent Mortality	Standard Error	# Chambers
A	4.9	(1.6)	8	0.0	(0.0)	8
B	4.4	(1.9)	8	0.0	(0.0)	8
C	1.7	(1.1)	8	0.8	(0.8)	8
D	1.7	(1.7)	8	0.0	(0.0)	8
E	4.1	(2.2)	8	0.8	(0.8)	8
F	3.6	(1.9)	8	2.5	(2.5)	8
G	3.5	(1.3)	8	0.0	(0.0)	8
H	-	-	-	0.0	(0.0)	8
I	1.8	(1.2)	8	0.0	(0.0)	8

Table 4-4. 30-Day Fish Mortality.

	Fathead Minnow			Channel Catfish		
Station	% Mortality	Standard Error	# Chambers	% Mortality	(SE)	# Chambers
A	12.5	(3.6)	8	0.0	(0.0)	8
B	7.7	(3.0)	8	0.0	(0.0)	8
C	0.9	(0.9)	8	0.0	(0.0)	8
D	3.3	(1.7)	8	1.7	(1.1)	8
E*	9.9	(2.5)	8	42.0	(15.0)	8
F*	10.8	(3.8)	8	44.8	(15.4)	8
G	5.2	(1.8)	8	1.7	(1.1)	8
H	-	-	-	0.0	(0.0)	8
I	4.3	(1.8)	8	0.8	(0.8)	8
		Neglecting unexplained mass mortality:				
E	NA			14.3	(5.0)	8
F	NA			18.3	(5.0)	8

To address channel catfish mortality at the three chambers at Station E and two chambers at Station F, the dead channel catfish and remaining live channel catfish were sent to the University of Nebraska Lincoln, Department of Veterinary and Biomedical Sciences, Veterinary Diagnostic Center for histopathological examination. Microscopic examination of tissue revealed no lesions or physical abnormalities that would be indicative of viral or bacterial infections. Gross examination only revealed a tenacious mucous adhered to the gills. Additionally, channel catfish were sent to the University of Montana for histopathological analysis. The histopathological analysis revealed no appreciable difference in the samples except for several parasitic worms in the intestine of one fish. The laboratory also reported seeing some *Henneguya* cysts on the gills and integument of some fish, but no great differences between the specimens.

To supplement these pathological examinations, more detailed studies were conducted in the vicinity of the high mortality fish chambers to help determine if ammonia or other water quality parameters might explain the mortalities. Water quality results, including conductivity, dissolved oxygen, ammonia, pH, and temperature, were not different from other chambers within the Station, and no other obvious sources of contamination were observed. Based on the laboratory and water quality evaluations, the mortalities at the specific fish chambers did not suggest a dose-response relationship (e.g., toxicity) caused by ammonia or other effluent-related contaminants. It is likely that a bacterial or fungal disease transmitted among the catfish in the same chamber, osmotic stress, or stress to channel catfish in confined spaces affected the mortality.

Figure 4-4. Distribution of Channel Catfish and Fathead Minnow Mortality at *In Situ* Station E.

SITE "E"

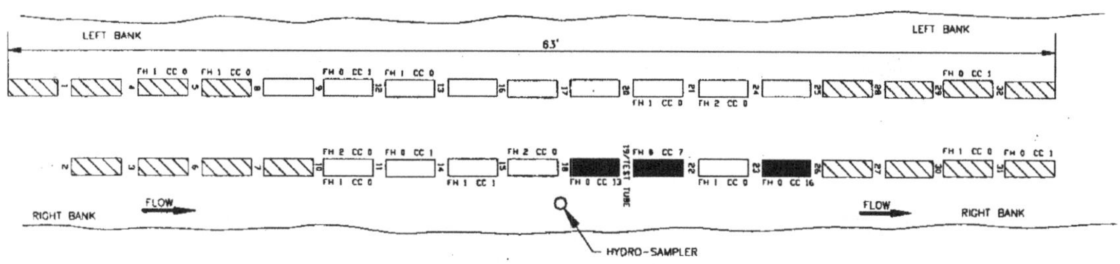

WATER QUALITY PARAMETERS IN SITU LOCATION E, 8 OCTOBER 1999, 0844 HOURS.					
POSTION	pH pH UNITS	CONDUCTIVITY UMHO/CM	DISSOLVED OXYGEN mg/L	TEMPERATURE DEGREES C	TOTAL NH3N mg/L
8	7.73	5060	6.36	18.7	10.31
10	7.71	4780	6.51	19.2	10.99
15	7.69	4780	7.09	19.2	11.01
16	7.72	5150	6.52	18.7	10.51
HYDROLAB	7.66	4640	6.47	19.4	11.32
19/TEST	7.67	4790	7.29	19.2	11.15
20	7.71	5260	6.46	18.5	10.47
23	7.64	4870	7.11	19	10.94
24	7.66	5210	6.11	18.3	10.41

DRAWING NAME: C:\ACAD2000\Drawing1.dwg
DATE:02/24/2000 TIME:15:53 DRAWN BY: sdolph

NOT TO SCALE

FIGURE: 2-4
DISTRIBUTION OF CHANNEL CATFISH AND FATHEAD MINNOW MORTALITY AT IN SITU LOCATION E, 21 SEPTEMBER – 21 OCTOBER 1999.

EA ENGINEERING, SCIENCE, AND TECHNOLOGY

Figure 4-5. Distribution of Channel Catfish and Fathead Minnow Mortality at *In Situ* Station F.

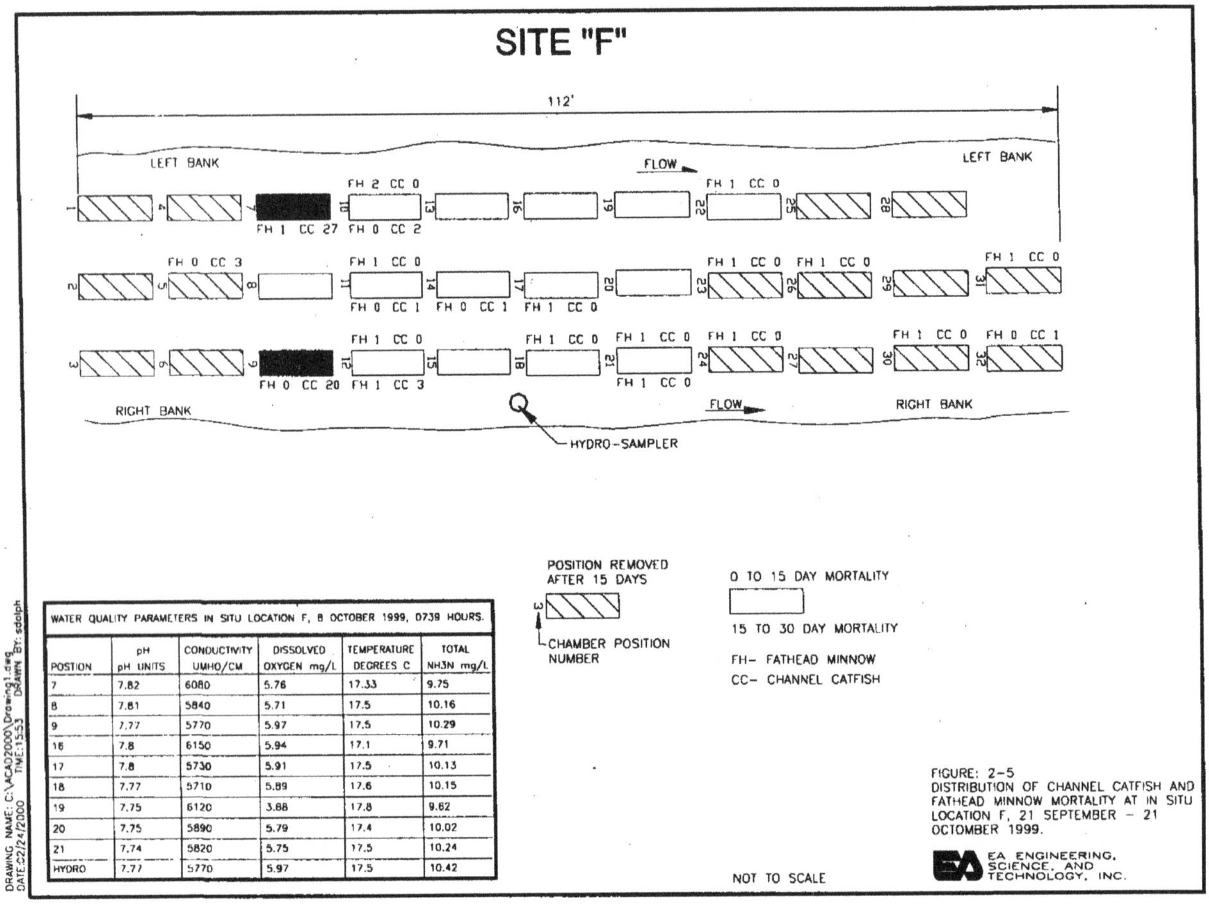

WATER QUALITY PARAMETERS IN SITU LOCATION F, 8 OCTOBER 1999, 0739 HOURS.

POSTION	pH pH UNITS	CONDUCTIVITY UMHO/CM	DISSOLVED OXYGEN mg/L	TEMPERATURE DEGREES C	TOTAL NH3N mg/L
7	7.82	6080	5.76	17.33	9.75
8	7.81	5840	5.71	17.5	10.16
9	7.77	5770	5.97	17.5	10.29
16	7.8	6150	5.94	17.1	9.71
17	7.8	5730	5.91	17.5	10.13
18	7.77	5710	5.89	17.6	10.15
19	7.75	6120	3.88	17.8	9.62
20	7.75	5890	5.79	17.4	10.02
21	7.74	5820	5.75	17.5	10.24
HYDRO	7.77	5770	5.97	17.5	10.42

4.2.6 Termination of *In Situ* Test

Channel catfish and fathead minnows were removed from the *in situ* fish chambers after 15 days of exposure and 30 days of exposure. Fish were removed at the 15-day duration in the event environmental impacts (such as high flow from a storm event) occurred that would destroy the *in situ* testing chambers prior to the end of the 30-day period. Fish were removed from the chambers at each Station in the same sequence that they had been deployed when the study was initiated. All fish were removed and placed in labeled one-gallon ziploc bags. A minimal amount of water was added to the bag to prevent water loss from the fish. The fish were then placed on ice in a cooler and transported to the laboratory, where all fish were measured and weighed (wet and dry) using the same procedures and equipment used to establish initial fish lengths and weights prior to initiation of *in situ* testing, as described in the Test Initiation section.

4.2.7 Fish Growth and Biomass Evaluation

Growth and biomass at each Station were used as the chronic sub-lethal endpoints for the *in situ* test. At the termination of the *in situ* test for the 15- and 30-day exposure periods, the surviving fish from each chamber were measured and weighed (wet and dry) as described in the Test Initiation section. The 15- and 30-day biomass data for channel catfish and fathead minnows are presented in Figures 4-6 and 4-7 and in Tables 4-5 and 4-6. Study period chloride, total ammonia, and un-ionized ammonia concentrations are also shown in Figures 4-6 and 4-7.

Figure 4-6. Summary of 15-Day Average Chamber Biomass.

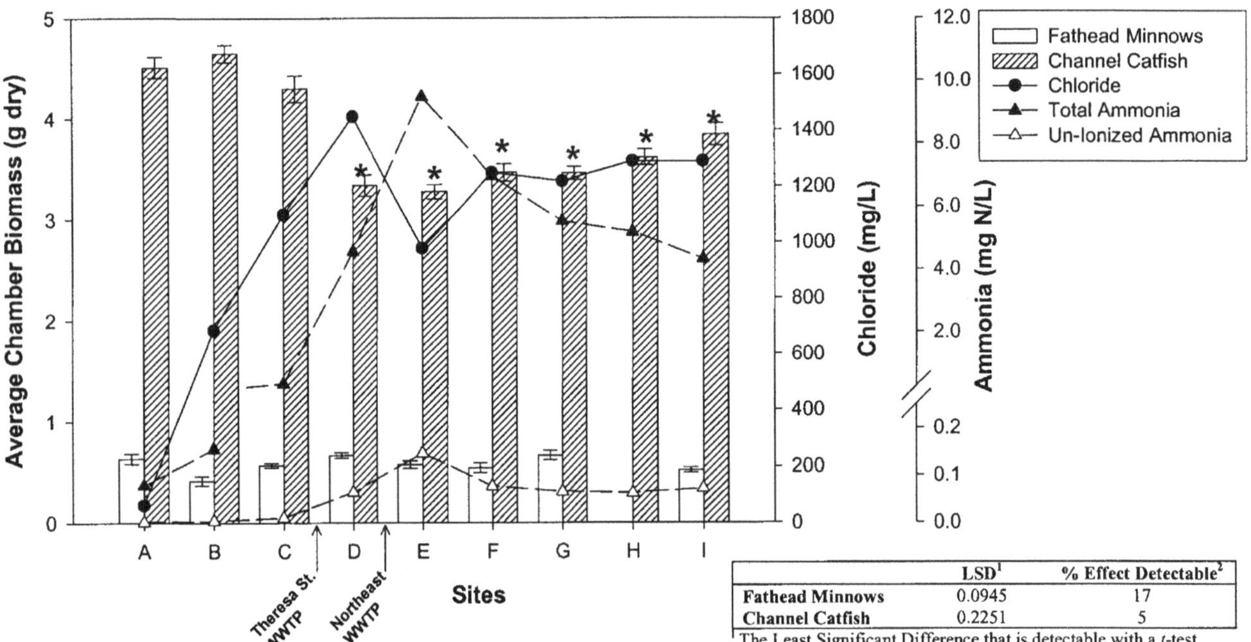

	LSD[1]	% Effect Detectable[2]
Fathead Minnows	0.0945	17
Channel Catfish	0.2251	5

[1] The Least Significant Difference that is detectable with a *t*-test.
[2] The 20% effect relative to Site C.

Average chamber biomass (±1 SE) after the 15-day *in situ* test. Asterisks (*) indicate the stations downstream of the WWTPs that show significant impairment relative to the control Site C, as determined by pairwise one-tailed *t*-tests (H0: $\overline{X}_C \leq \overline{X}_i$; $\alpha = 0.05$). Average chloride and both total and un-ionized ammonia concentrations are superimposed.

Figure 4-7. Summary of 30-Day Average Chamber Biomass.

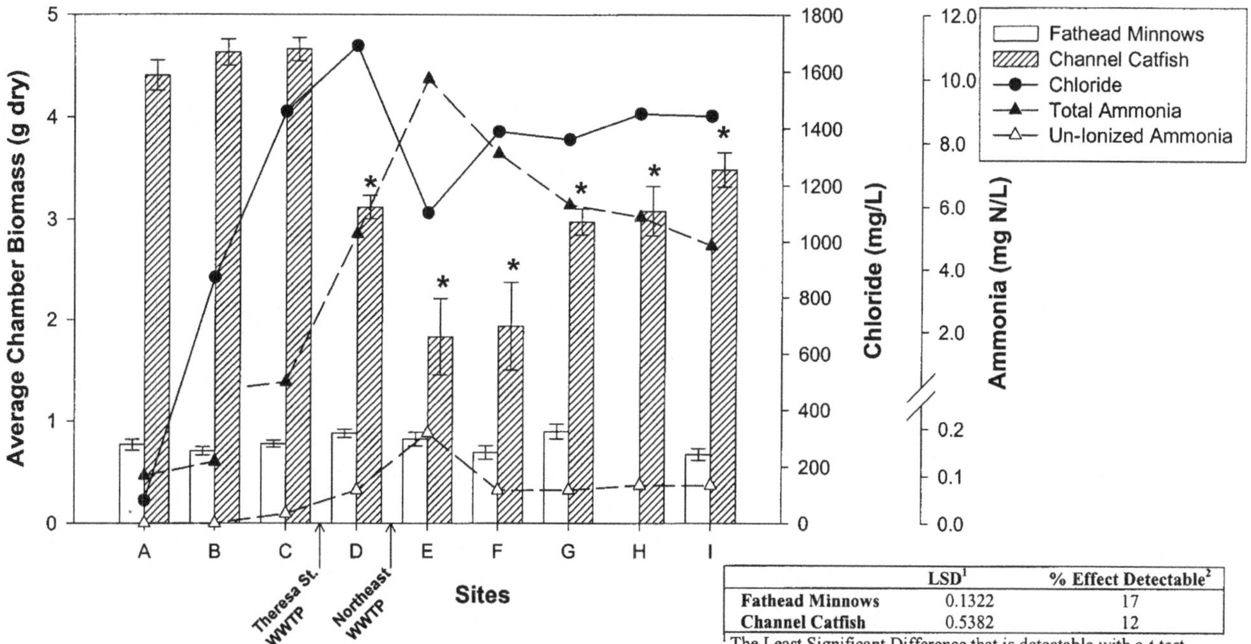

	LSD[1]	% Effect Detectable[2]
Fathead Minnows	0.1322	17
Channel Catfish	0.5382	12

[1] The Least Significant Difference that is detectable with a *t*-test.
[2] The 20% effect relative to Site C.

Average chamber biomass (±1 SE) after the 30-day *in situ* experiment. Asterisks (*) indicate the stations downstream of the WWTPs that show significant impairment relative to the control Site C, as determined by pairwise one-tailed *t*-tests (H_0: $\overline{X}_C \le \overline{X}_i$; $\alpha = 0.05$). Average chloride and both total and un-ionized ammonia concentrations are superimposed.

4.2.7.1 15-Day Exposure Results for Growth

Although the U.S. EPA, in the ammonia criteria documents (1998, 1999), expresses a preference for chronic test exposures of 30 days or longer, the 15-day growth results were still evaluated. The 15-day test results provided a mid-point check for the *in situ* testing, plus provided *in situ* testing results in the event environmental impacts (high flow, etc.) did not allow the 30-day test to be completed. A summary of the 15-day test results is as follows:

- For fathead minnows there was substantial growth during the 15-day exposure period that varied between Stations, but does not appear to be related to concentrations of ammonia. As shown in Figure 4-8, the highest percent growth (relative to initial weights) was at Stations G (130.7%t) and D (127.6%), and at upstream station A (120.3%). The lowest percent growth was at upstream Station B (45.1%).
- For channel catfish, the percent growth was lower than that of the fathead minnows and the highest relative growth occurred at the three upstream stations (Figure 4-8 and Table 4-5). Although the mean chloride concentrations at Stations A, B and C, which are above the WWTPs, ranged from approximately 60 to1,100 mg/L, the 15-day growth data for the channel catfish do not substantively differ for these three stations.

Table 4-5. 15-Day Dry Weights and Percent Growth for Fathead Minnows and Channel Catfish.

	Fathead Minnow					Channel Catfish				
Site	Dry Weight (SE)		% Growth (SE)		n	Dry Weight (SE)		% Growth (SE)		n
Initial	0.0203	(0.0017)	NA	NA	20	0.1865	(0.0083)	NA	NA	20
A	0.0448	(0.0023)	120.29	(11.46)	113	0.3008	(0.0082)	61.28	(4.42)	120
B	0.0295	(0.0013)	45.10	(6.53)	112	0.3051	(0.0085)	63.57	(4.58)	122
C	0.0394	(0.0017)	93.95	(8.14)	115	0.2889	(0.0078)	54.88	(4.16)	119
D	0.0463	(0.0018)	127.62	(8.98)	115	0.2228	(0.0065)	19.47	(3.49)	120
E	0.0398	(0.0022)	95.67	(10.66)	115	0.2204	(0.0068)	18.17	(3.66)	119
F	0.0396	(0.0021)	94.52	(10.38)	109	0.2391	(0.0066)	28.21	(3.52)	116
G	0.0469	(0.0020)	130.74	(9.93)	113	0.2308	(0.0067)	23.72	(3.62)	120
H	-	-	-	-	-	0.2432	(0.0068)	30.36	(3.66)	119
I	0.0371	(0.0015)	82.54	(7.41)	111	0.2584	(0.0067)	38.55	(3.58)	119

SE = standard error of the mean

Figure 4-8. Summary of 15-Day Fish Percent Growth.

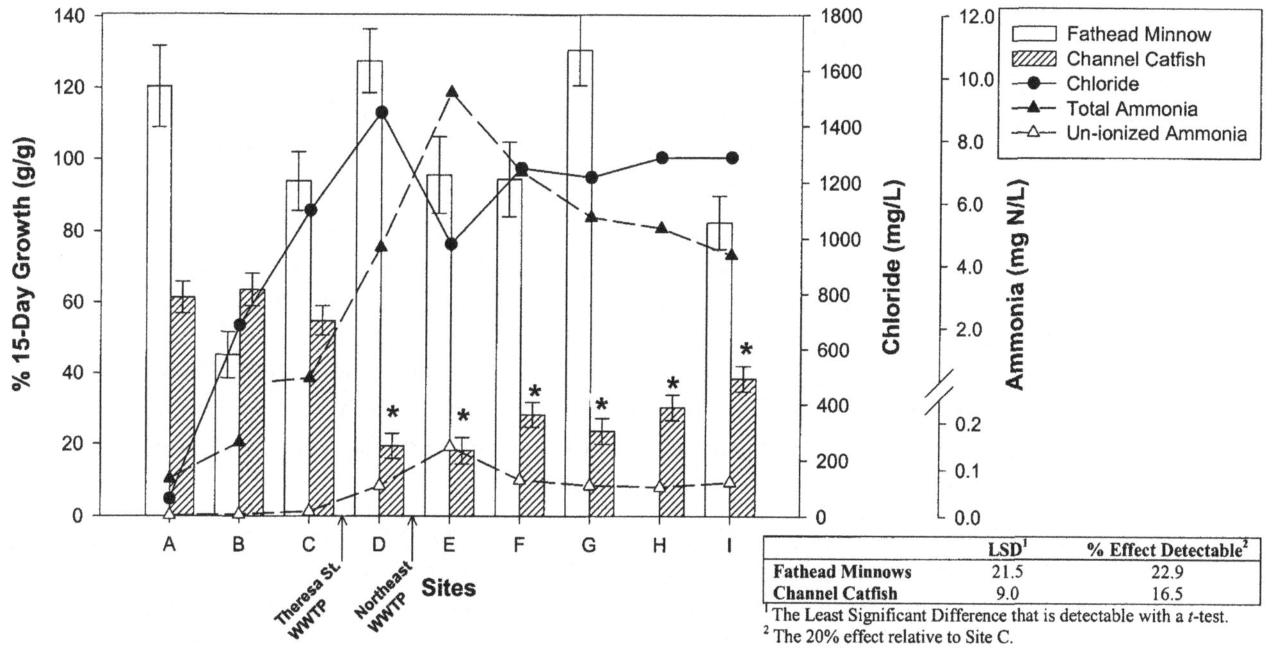

	LSD[1]	% Effect Detectable[2]
Fathead Minnows	21.5	22.9
Channel Catfish	9.0	16.5

[1] The Least Significant Difference that is detectable with a *t*-test.
[2] The 20% effect relative to Site C.

Mean dry weights (±1 SE) after the 15-day *in situ* experiment. Asterisks (*) indicate the stations downstream of the WWTPs that show significant impairment relative to the control Site C, as determined by pairwise one-tailed *t*-tests (H_0: $\overline{X}_C \leq \overline{X}_i$; $\alpha = 0.05$). The average chloride and both total and un-ionized ammonia concentrations are superimposed.

4.2.7.2 30-Day Exposure Results for Growth

Growth results for the 30-day exposure test for fathead minnows and channel catfish are summarized in Figure 4-9 and in Table 4-6. A summary of the results is as follows:

- For fathead minnows there was substantial growth during the total 30-day exposure period—as well as between 15 and 30 days. As shown in Figure 4-9, the highest percent growth (relative to initial weights) over the 30 days was at downstream Stations G (216%), D (203.4%), and E (197%), and at upstream Station A (182.7%). The lowest percent growth was at the furthest downstream Station, Station I (153.4%). Based on the 30-day exposure data, the growth of fathead minnows varied between stations but does not appear to be related to concentrations of ammonia. Mean chloride concentrations at upstream Stations A, B, and C (above the WWTPs) ranged from approximately 80 to 1,460 mg/L; the 30-day growth data for the fathead minnows do not substantively differ for these three stations.
- For channel catfish the percent growth was lower than fathead minnows, and little additional growth occurred during the 15- to 30-day timeframe (Tables 4-5 and 4-6). The highest relative growth occurred at the three upstream stations with percent growth values of 57.4%, 69.7%, and 63.8%t at Stations A, B, and C, respectively (Figure 4-9 and Tables 4-5 and 4-6). Further, as shown in Figure 4-9, growth at Stations D through I was significantly lower than at upstream stations A, B, and C. As in the 15-day catfish data, although the mean chloride concentrations at upstream Stations A, B, and C ranged from approximately 80 to 1,460 mg/L, the 30-day growth data for the channel catfish do not substantively differ for these three upstream stations.

Table 4-6. 30-Day Dry Weights and Percent Growth for Fathead Minnows and Channel Catfish.

	Fathead Minnow					Channel Catfish				
Site	Dry Weight (SE)		% Growth (SE)		n	Dry Weight (SE)		% Growth (SE)		n
Initial	0.0203	(0.0017)	NA	NA	20	0.1865	(0.0083)	NA	NA	20
A	0.0575	(0.0042)	182.66	(20.50)	107	0.2937	(0.0078)	57.44	(4.19)	120
B	0.0534	(0.0045)	162.65	(21.93)	106	0.3166	(0.0093)	69.73	(4.99)	117
C	0.0542	(0.0020)	166.53	(9.82)	115	0.3056	(0.0075)	63.83	(4.01)	122
D	0.0617	(0.0024)	203.44	(11.69)	114	0.2115	(0.0065)	13.40	(3.47)	118
E	0.0604	(0.0029)	197.01	(14.12)	109	0.1885	(0.0080)	1.05	(4.27)	68
F	0.0518	(0.0023)	154.94	(11.21)	107	0.2052	(0.0083)	10.03	(4.47)	66
G	0.0643	(0.0034)	215.99	(16.64)	112	0.2107	(0.0059)	12.95	(3.17)	113
H	-	-	-	-	-	0.2201	(0.0062)	17.99	(3.34)	112
I	0.0515	(0.0022)	153.43	(10.63)	105	0.2386	(0.0067)	27.92	(3.62)	117

SE = standard error of the mean

Figure 4-9. Summary of 30-Day Fish Percent Growth.

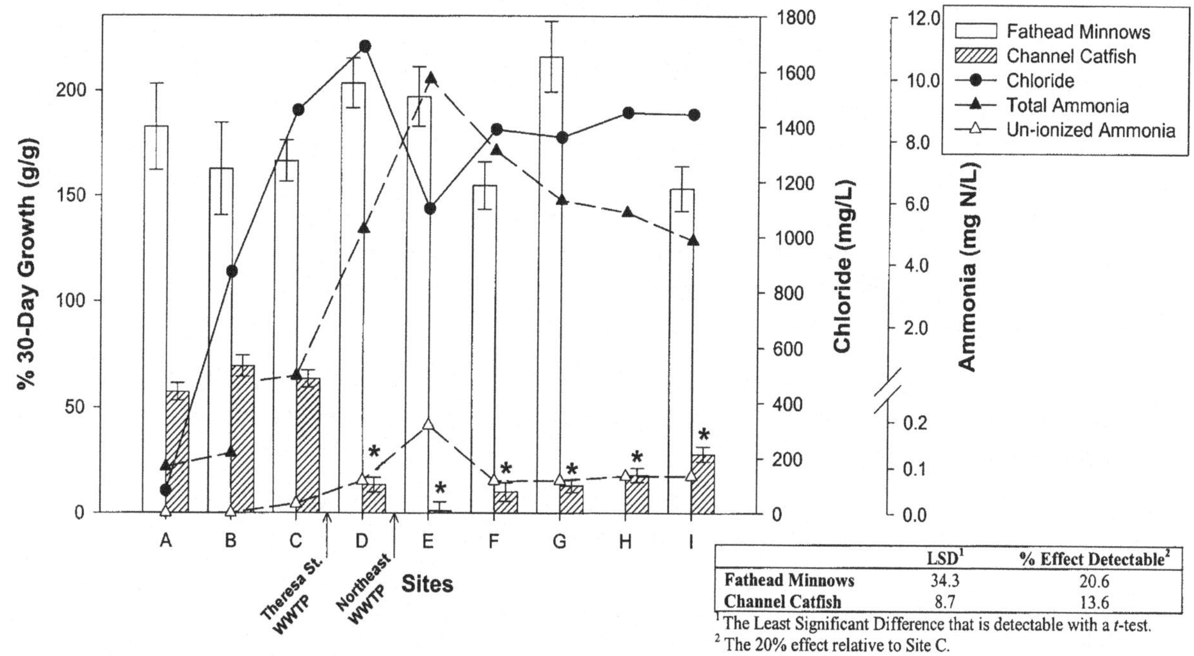

	LSD[1]	% Effect Detectable[2]
Fathead Minnows	34.3	20.6
Channel Catfish	8.7	13.6

[1] The Least Significant Difference that is detectable with a *t*-test.
[2] The 20% effect relative to Site C.

Mean dry weights (±1 SE) after the 30-day *in situ* experiment. Asterisks (*) indicate the stations downstream of the WWTPs that show significant impairment relative to the control Site C, as determined by pairwise one-tailed *t*-tests (H_0: $\overline{X}_C \leq \overline{X}_i$; α = 0.05). The average chloride and both total and un-ionized ammonia concentrations are superimposed.

4.2.8 Ambient Temperature During the *In Situ* Study

Ambient Salt Creek temperature data for the 30-day *in situ* study period are presented in Figure 4-10. Because Salt Creek is shallow during low flow periods, diurnal temperature changes can be substantial, and the affects of temperature need to be considered in developing conclusions from the *in situ* testing study. The introduction of wastewater from the Theresa Street and Northeast WWTPs increases the temperature of Salt Creek during cooler periods of the year. As shown in Figure 4-10, the increase from Theresa was approximately 1.5° Celsius, and the increase from the combined Theresa Street and Northeast WWTP discharges was approximately 3° Celsius based upon the 30-day average values. This ambient temperature increase continues downstream but is dampened with distance.

Ambient temperature has been shown to affect the growth of aquatic fish and invertebrates. *In situ* study fish growth and biomass data were evaluated with and without temperature correction.

Figure 4-10. 1-15 Day, 15-30 Day, and 1-30 Day Maximum, 75th Percentile, Average 25th Percentile, and Minimum Temperatures During the Salt Creek *In Situ* Study.

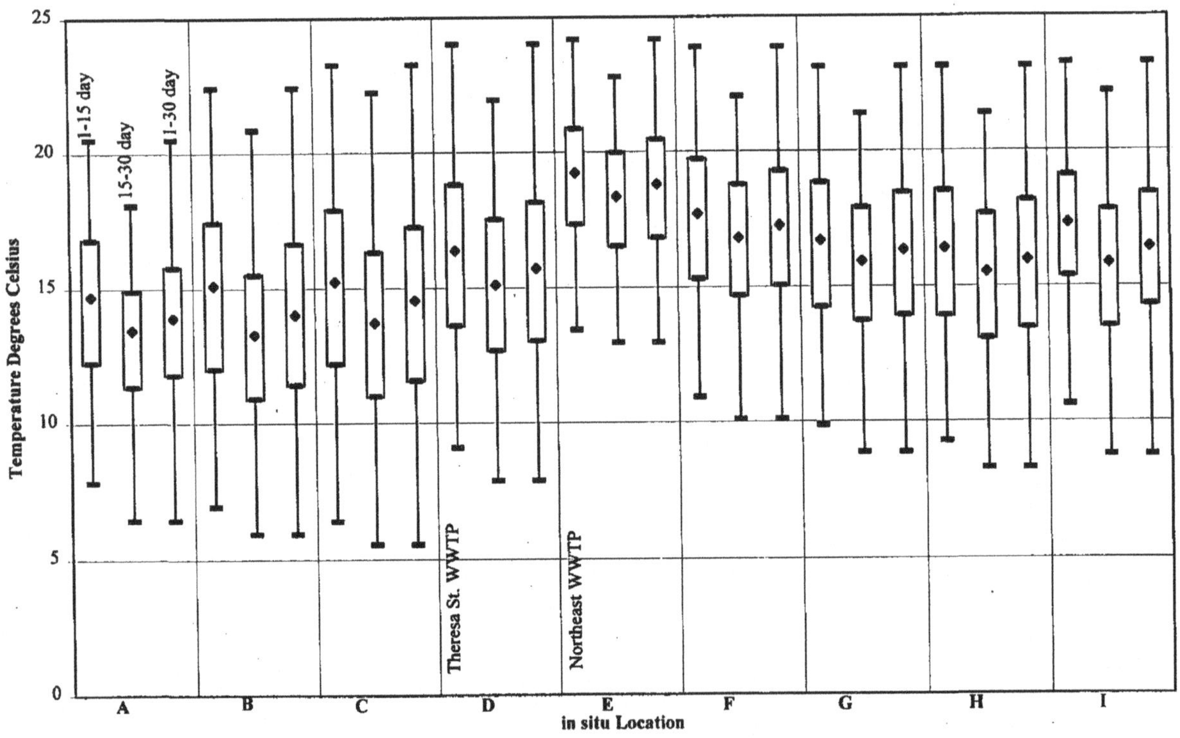

4.3 *In Situ* Toxicity Testing Results

The primary goal of the *in situ* toxicity testing program was to calculate 30-day IC20 values for fathead minnows and channel catfish, and use those data in combination with the other studies to determine a weight-of-evidence based site-specific water quality criterion for ammonia, incorporating the following:

- 30-day exposures are close to the minimum test duration used for chronic fish in the ammonia criteria update. As stated in the U.S. EPA 1999 Update of Ambient Water Quality Criteria for Ammonia (U.S. EPA, 1999), 70% (14/20) of the chronic fish tests had exposure durations of between 28 and 32 days. Two of the remaining six tests were life cycle tests and the other four were 42-73 day rainbow trout tests.
- Biomass is the preferred endpoint (versus growth) because it combines the effects of survival and growth for each exposure concentration, and "*because growth effects on juvenile fish are sometimes transient, so that short chronic tests with such tests might overestimate risk*" (U.S. EPA, 1999). For this specific reason, U.S. EPA (1999) chose to not use growth effect data for channel catfish for criteria derivation (p. 71).

The calculated *in situ* IC20 values for total ammonia and un-ionized ammonia are presented in Tables 4-7 and 4-8 below, respectively. It is important to recognize that the IC20 values reflect the actual water quality conditions of Salt Creek during the 30-day exposure period (e.g., the in-stream variability of ammonia, pH, temperature, chloride, and other background water quality characteristics).

Table 4-7. Calculated IC20 Values for Total Ammonia.

	IC20 ± Standard Error (mg/L)			
	Fathead Minnows		Channel Catfish	
	15-Day	30-Day	15-Day	30-Day
% Mortality	na	na	na	6.75 ±0.37
% Growth	na	na	3.02 ±0.68	1.94 ±0.29
Total Biomass	> 9.5	> 9.98	7.33 ±1.61	3.85 ±0.48

Note: Standard errors for the IC20 were computed from the bootstrap method with 1,000 resamples.

Table 4-8. Calculated IC20 Values for Un-ionized Ammonia.

	IC20 ± Standard Error (mg/L)			
	Fathead Minnows		Channel Catfish	
	15-Day	30-Day	15-Day	30-Day
% Mortality	na	na	na	0.10 ±0.02
% Growth	na	na	0.03 ±0.00	0.03 ±0.00
Total Biomass	na	na	0.09 ±0.02	0.05 ±0.00

Note: Standard errors for the IC20 were computed from the bootstrap method with 1,000 resamples.

4.3.1 Channel Catfish Result

The results of the 15- and 30-day *in situ* exposures are summarized in Figures 4-11 and 4-12, respectively. The 30-day IC20 value for the channel catfish was calculated to be 3.85 mg/L based on total biomass (Figure 4-12). The corresponding 15-day IC20 value was calculated as 7.33 mg/L (Figure 4-11), which is within the relative range expected for a shorter duration test. These values were not "adjusted" in any way to address the unexplained mortalities at Stations E and F, as described above. The calculated IC20 is not affected by the inclusion of the observed mass mortalities, since they occurred only at Stations E and F, and (as shown in Figure 4-12) the IC20 value is calculated based on the data from Stations C, I, H, and G.

Key Result for 30-Day Test: Channel catfish IC20 = 3.85 mg/L

4.3.2 Fathead Minnow Results

The 30-day IC20 values for the fathead minnow are greater than 9.98 mg/L, since there was less than a 20% decrease in biomass at downstream stations compared to the control (Station C) at the highest 30-day average in-stream exposure concentration. The lack of a dose-response to the fathead minnows is clearly shown in Figures 4-11 and 4-12, but significant fathead minnow growth was evident in both the 15- and 30-day exposure periods. The lack of a dose-response was unexpected given the literature values presented in the U.S. EPA's (1998) ammonia criteria update, which were used for the design of the *in situ* toxicity testing program. One explanation for the fathead minnow result identified in U.S. EPA's 1999 Update of Ambient Water Quality Criteria for Ammonia (U.S. EPA, 1999) is that juvenile fathead minnows have been shown to be less sensitive than early life stages by a factor of approximately three.

Key Result for 30-Day Test: Fathead minnow IC20 = >9.98 mg/L

Figure 4-11. 15-Day Total Chamber Biomass Versus Mean Total Ammonia Concentration.

Dose-response relationship between total chamber biomass and mean total ammonia concentration during the 15-day *in situ* experiment. The mean total chamber biomass (solid line) is used to calculate the IC_{20}. The dashed line indicates the IC_{20} for channel catfish. For fathead minnows, a 20% effect relative to the reference site C was not observed.

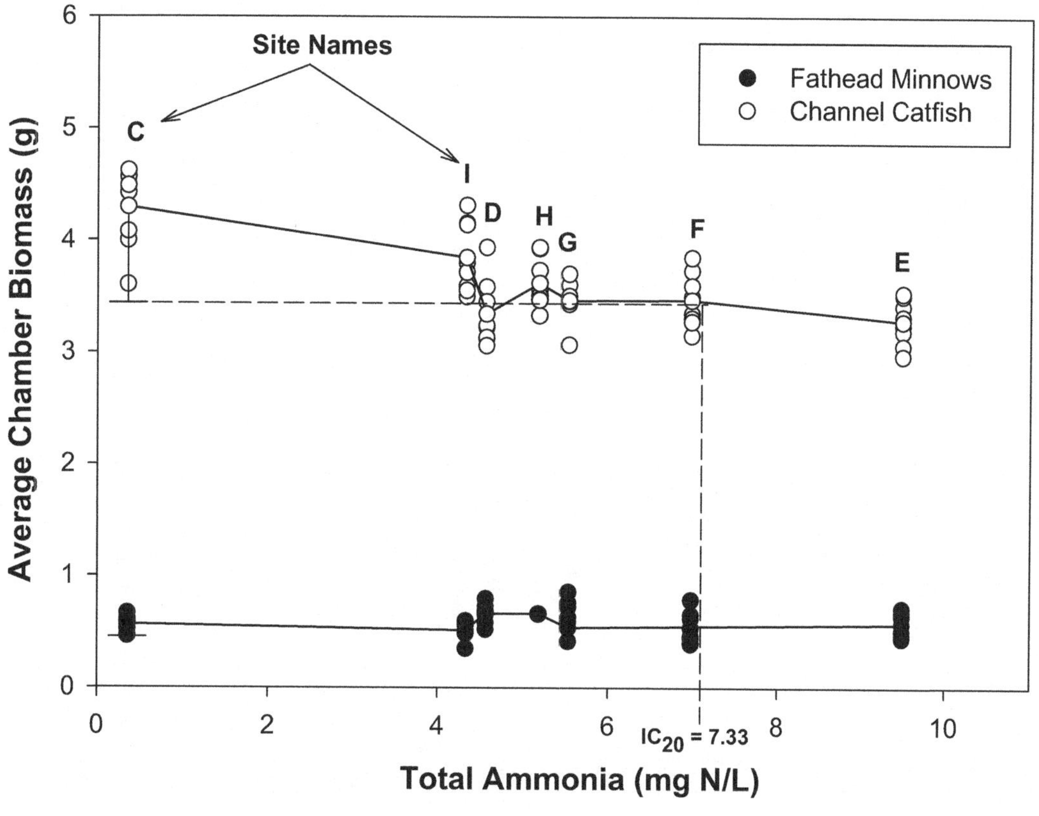

Figure 4-12. 30-Day Total Chamber Biomass Versus Mean Total Ammonia Concentration.

Dose-response relationship between total chamber biomass and mean total ammonia concentration during the 30-day *in situ* experiment. The mean total chamber biomass (solid line) is used to calculate the IC_{20}. The dashed line indicates the IC_{20} for channel catfish. For fathead minnows, a 20% effect relative to the reference site C was not observed.

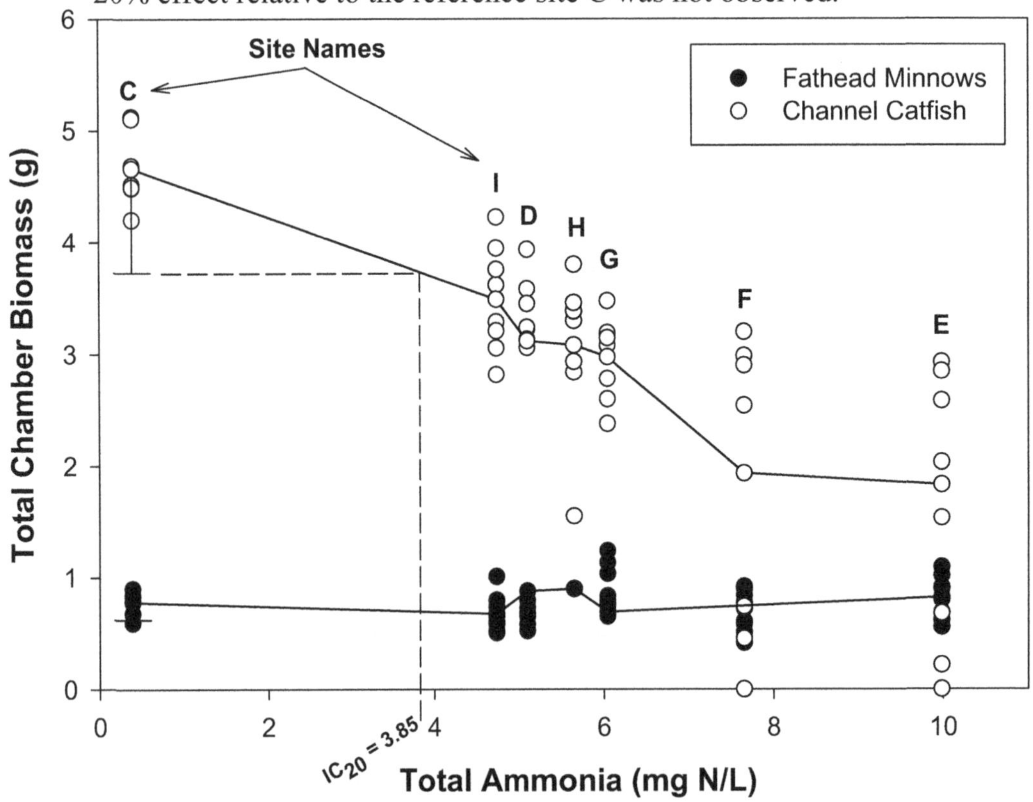

4.3.3 Statistical Associations Between Water Quality Variables and Biological Endpoints

Pearson product-moment correlation coefficients (r) were computed to obtain the degree of statistical association between biological endpoints and selected water quality parameters from the *in situ* toxicity testing. The matrix of correlation coefficients for the 30-day channel catfish results (Table 4-9 and Table 4-10) shows a high degree of inverse association between total biomass and total ammonia ($r = -0.96$, $p < 0.001$). There were no significant correlations observed in the 30-day fathead minnow data.

Table 4-9. Channel Catfish Correlation Coefficients Between Water Quality Parameters and Toxicity Metrics for 30-Day Test.

	Total Ammonia	Un-ionized Ammonia	Conductivity	Chlorides	Temperature	pH
Dry Weight	0.32 (0.4460)	0.33 (0.4127)	-0.04 (0.9335)	0.04 (0.9224)	0.26 (0.5262)	-0.15 (0.7165)
%Mortality	0.18 (0.6625)	0.12 (0.8096)	-0.70 (0.0546)	-0.72 (0.0439)	0.19 (0.6479)	-0.56 (0.1484)
%Growth	0.32 (0.4460)	0.33 (0.4127)	-0.04 (0.9335)	0.04 (0.9224)	0.26 (0.5262)	-0.15 (0.7165)
Total Biomass	0.25 (0.5517)	0.26 (0.5111)	0.11 (0.8037)	0.19 (0.6465)	0.21 (0.6195)	-0.04 (0.9333)

Note: Pearson Product-Moment Correlation Coefficients and their probabilities of significance for the H_0: r = 0

Table 4-10. Fathead Minnow Correlation Coefficients Between Water Quality Parameters and Toxicity Metrics for 30-Day Test.

	Total Ammonia	Un-ionized Ammonia	Conductivity	Chlorides	Temperature	pH
Dry Weight	-0.96 (0.0000)	-0.84 (0.0055)	-0.50 (0.1719)	-0.50 (0.1694)	-0.89 (0.0013)	0.68 (0.0441)
%Mortality	0.68 (0.0449)	0.58 (0.1116)	0.08 (0.8348)	0.07 (0.8628)	0.75 (0.0190)	-0.60 (0.0892)
%Growth	-0.96 (0.0000)	-0.84 (0.0055)	-0.50 (0.1719)	-0.50 (0.1694)	-0.89 (0.0013)	0.68 (0.0441)
Total Biomass	-0.97 (0.0000)	-0.85 (0.0050)	-0.37 (0.3271)	-0.36 (0.3369)	-0.96 (0.0001)	0.74 (0.0213)

Note: Pearson Product-Moment Correlation Coefficients and their probabilities of significance for the H_0: r = 0.

CHAPTER 5.0

AMMONIA CRITERIA DEVELOPMENT AND REGULATORY PROCEDURES

The U.S. EPA 1991 Technical Support Document for Water Quality-based Toxics Control states the following:

> *"Section 101(a) of the Clean Water Act states: "The objective of this Act is to restore and maintain the chemical, physical and biological integrity of the Nation's waters." Taken together, chemical, physical and biological integrity define the overall ecological integrity of an aquatic ecosystem. Regulatory agencies should strive to fully integrate all three approaches since each has its respective capabilities and limitations".*

The NDEQ also states provisions in their *Title 117 – Nebraska Surface Water Quality Standards* that acceptable conditions occur for developing site-specific criteria. Under Title 117 Chapter 4 – *Standards for Water Quality*, sub-section 003.02A6b, NDEQ allows for site-specific analysis for the modification of existing water quality criteria conducted in accordance with various procedures that are described in the *Water Quality Standards Handbook* (U.S. EPA 1983). One of the accepted procedures is listed under Title 117 sub-section 003.02A6B(4), which states the following:

> *"Other scientifically defensible procedures such as relevant aquatic field studies, laboratory tests, or available scientific literature".*

The SCWQS followed this "other scientifically defensible procedures" approach and was the basis for selecting site-specific chronic total ammonia water quality criteria that are protective of the designated uses of Salt Creek Segment LP2-20000. In calculating site-specific total ammonia criteria, the City implemented a weight-of-evidence approach by using the numeric 30-day test period *in situ* toxicity study results, and using the extensive site-specific biological assessment results as supporting or corroborating evidence. This approach was developed under the scientific guidance of the WERF Peer Review Panel. Reasons for using both *in situ* toxicity testing and biological assessment results in a weight-of-evidence approach lie in the particular strengths of each measure to assess aquatic life protection for ammonia. *In situ* tests specifically evaluate ammonia impacts to fish growth and survival as chronic sub-lethal responses and are a real time measure of potential impacts. Biological assessments provide a measurement of the potential impairment to the largest number of species in Salt Creek for the longest period of exposure. The data express the integration of impacts over time to the aquatic community and, by considering the results together, a protective total ammonia criteria was developed for Salt Creek Segment LP2-20000.

5.1 Development of Total Ammonia Chronic Criteria

As discussed in Chapter 4.0, the fathead minnow *in situ* toxicity testing results did not measure a 20% effect (IC20) relative to the reference station. The 30-day IC20 values for the fathead minnow were greater than 9.98 mg /L, since there was less than a 20% decrease in biomass at downstream stations, compared to the reference station. With this result it was decided, in conjunction with NDEQ and the WERF Peer Review Panel, that only the channel catfish IC20 concentrations (3.85 mg/L) would be used to determine a chronic ammonia water quality criterion. This approach is conservative, since the channel catfish IC20 is substantially less than the greater than 9.98 mg/L value for fathead minnows.

Salt Creek *in situ* toxicity tests were conducted during the month of October, and the testing results, both biological and chemical, are most representative of that time of year. In discussions with NDEQ Surface Water Section, Water Quality Division staff and the WERF Peer Review Panel, seasonality effects (pH and temperature) on ammonia needed to be accounted for to ensure the final site-specific chronic ammonia criterion was protective during NDEQ's two regulatory seasons—summer (March through October) and winter (November through February). To account for seasonal effects, NDEQ's *Procedures for Developing Point Source Total Maximum Daily Loads* guidance, and U.S. EPA's 1999 *Update of Ambient Water Quality Criteria for Ammonia* document were used to adjust *in situ* testing results.

5.1.1 *In Situ* Testing pH and Temperature

As discussed in Chapter 4.0, Salt Creek water quality was monitored daily during the 30-day *in situ* study period. Hydrolab™ units were used at each station to measure water quality parameters, including pH and temperature, which were collected at 15 minute intervals for a total of approximately 2,880 measurements for each parameter. As required under NDEQ's guidance (Title 117), the critical month median pH value and 75th percentile temperature value are used for the calculation of chronic total ammonia, or un-ionized ammonia, water quality criteria. The median month pH value and 75th percentile temperature value for all stations below the WWTP discharge (Stations D through I) were 7.7 s.u. and 17° Celsius, respectively.

5.1.2 Seasonal Correction Factor

The channel catfish IC20 concentration of 3.85 mg/L was adjusted to account for pH and temperature seasonal effects on ammonia. To account for seasonal effects, the channel catfish IC20 was adjusted to the U.S. EPA 1999 *Update of Ambient Water Quality Criteria for Ammonia*, which was adopted into Title 117 by NDEQ as follows:

- Adjustment to U.S. EPA Criteria
    - Channel catfish study value pH of 7.7 s.u. and temperature of 17° Celsius equates to a 3.0 mg/L total ammonia summer season criterion based on the U.S. EPA 1999 *Update of Ambient Water Quality Criteria for Ammonia.*
- Adjustment to Seasons
    - Summer Season

- Median Salt Creek pH and temperature for critical summer season month of August (per NDEQ requirement) is 8.1 s.u. and 25.5° Celsius, which equates to a total ammonia criterion of 1.07 mg/L based on the U.S. EPA 1999 *Update of Ambient Water Quality Criteria for Ammonia.*
- The reduction from 3.0 mg/L to 1.07 mg/L is 64%, or a 0.36 multiplier applied to the 3.85 mg/L IC20 value.
- The resulting summer total ammonia criterion is 1.4 mg/L (Early Life Stage Present).

- Winter Season
  - Median Salt Creek pH and temperature for critical winter season month of January (per NDEQ requirement) is 7.8 s.u. and 3.8° Celsius, which equates to a total ammonia criterion of 6.96 mg/L based on the U.S. EPA 1999 *Update of Ambient Water Quality Criteria for Ammonia.*
    - The increase from 3.85 mg/L to 6.96 mg/L is 181% or a 1.81 multiplier applied to the 3.85 mg/L IC20 value.
    - The resulting winter total ammonia criterion is 6.7 mg/L (Early Life Stage Absent).

5.1.3 Increase in Chronic Ammonia Criteria from *In Situ* Toxicity Tests

The increase in the chronic ammonia criteria based on the *in situ* toxicity tests is expressed as the difference in the criteria from the U.S. EPA 1999 *Update of Ambient Water Quality Criteria for Ammonia* document and the *in situ* IC20 value adjusted to the U.S. EPA criteria, as explained above. Applying this ratio, the *in situ* toxicity tests increased the site-specific ammonia chronic criteria by 28%, or 3.85 mg/L divided by 3.0 mg/L.

CHAPTER 6.0

CONCLUSIONS AND THE COLLABORATIVE REGULATORY AND TECHNICAL PROCESS

Under Title 117 *Nebraska Surface Water Quality Standards*, a chronic 30-day average chronic criterion for ammonia is calculated using a Criteria Continuous Concentration (CCC) and site-specific pH and temperature. The CCC is calculated using site-specific pH and temperature and a constant that represents a percentage of the lowest fish Genus Mean Chronic Value (GMCV). For Warmwater Class A streams, such as Salt Creek Segment LP2-20000, the constant in the CCC calculation is 0.854. Based on the final site-specific chronic ammonia criteria for Segment LP2-20000, a revised constant of 1.097 is used in the calculation of ammonia CCC. The difference between these two constants is an approximate 28% increase. The Segment LP2-20000 site-specific chronic ammonia criterion was proposed to the Nebraska Environmental Quality Council by NDEQ in late 2000, where it was approved for final adoption into Title 117.

The City of Lincoln worked closely with NDEQ, U.S. EPA Region VII, and the WERF Peer Review Panel during the SCWQS and development of the final, site-specific *in situ*-based, chronic ammonia criterion. The project was an example of successful and professional collaboration effort between a regulated entity and regulators, plus the integration of technical recommendations from the nationally recognized group of professionals which formed the WERF Peer Review Panel. The WERF Peer Review Panel's review of *in situ* study plans, review of final results, and endorsement of the *in situ* methodology ultimately led to the development of a defensible technical approach to developing site-specific chronic ammonia criteria for Salt Creek that could be adopted under state regulation with federal approval by U.S. EPA.

REFERENCES

A technique for assessing stream habitat structure. In Proc. Of Riparian Ecosystems. Barbour, M.T. and J.B. Stribling. Washington, D.C: National Association of Conservation Districts. 1994.

Salt Creek Water Quality Studies Demonstration Project Work Plan. June 4, 1998. City of Lincoln, June 4, 1998. Prepared for the City of Lincoln by Brown and Caldwell, EA Engineering, Science and Technology, and Kelly and Weaver.

Evaluating Whole Effluent Toxicity Testing as an Indicator of Instream Biological Condition. Diamond, J., et al. Water Environment Research Foundation, 635 Slaters Lane, Alexandria, VA, 22314-1177. 1999.

Stream Hydrology: An Introduction for Ecologists. Gordon, N.D., T.A. McMahon, and B.L. Finlayson. Centre for Environmental Applied Hydrology. University of Melbourne. John Wiley & Sons Ltd., Baffins Lane, Chichester, West Sussex. PO19 1UD, England. 1992.

Nebraska Stream Classification Study. Surface Water Section, Water Quality Division, Nebraska Department of Environmental Control, Ken Bazata, Report Coordinator. P.O. Box 98922, 301 Centennial Mall South, Lincoln, NE 68509-8922. 1991.

Nebraska Stream Classification Study. Draft Metric Revision for Nebraska Stream Classification Study (Draft). Surface Water Section, Water Quality Division, Nebraska Department of Environmental Control, Ken Bazata, Report Coordinator. P.O. Box 98922, 301 Centennial Mall South, Lincoln, NE 68509-8922. 1995.

Aquatic Evaluation and Use Attainability of Salt Creek, Nebraska. NDEQ, 1982. Nebraska Department of Environmental Quality. 1982.

Rapid Bio-assessment Protocols for Use in Streams and Rivers; Benthic Macroinvertebrates and Fish. Plafkin, et al. EPA/444/4-89-001. U.S. Environmental Protection Agency, Assessment and Watershed Protection Division, 401 M. Street, S.W., Washington, D.C. 20460. 1989.

SCWQSR, 1996 Vol. 1 of 2. Salt Creek Water Quality Studies Report, Review and Comment Edition, Volume 1 of 2. February 16, 1996. Prepared for the City of Lincoln.

Site-Specific Chronic Ammonia Criteria Final Technical Report. April 3, 2000. Revised June 16, 2000.SCWQS, 2000. City of Lincoln, Nebraska, Salt Creek Water Quality Studies.

Biological Criteria: Technical Guidance for Streams and Small Rivers, Revised Ed. EPA 822-B-94-001. U.S. Environmental Protection Agency. 1996. U.S. Environmental Protection Agency, Office of Science and Technology. Washington, D.C. May 1996.

Update of Ambient Water Quality Criteria for Ammonia. Document #EPA 822-R-98-008. U.S. Environmental Protection Agency. 1998. Office of Water, Washington, D.C.

1999 Update of Ambient Water Quality Criteria for Ammonia. Document #EPA 822-R-99-014. 1999. U.S. Environmental Protection Agency. Office of Water, Washington, D.C. Issued 22 December 1999.

Rapid Bio-assessment Protocols; 2nd ed. U.S. Environmental Protection Agency. Washington, D.C. 20460. 1999.

IWA Publishing's authorised EU representative for General Product Safety Regulations is Diane D'Arras, 15 rue Duret, 75116 Paris, France, e-mail: safety@iwap.co.uk.

Printed and bound by CPI Group (UK) Ltd, Croydon, CR0 4YY
17/09/2026
02218799-0010